HORTICULTURAL ENGINEERING TECHNOLOGY
FIELD MACHINERY

Other Related Books

Ayurvedic Medicinal Plants of India, Vol. 1-2 (Set)	Bhutiya, R.K.
Bacterial and Viral Disease and their Management in Horticultural Crops	Reddy, P.P.
Bamboo Flowering and Rodent Outbreaks	Sridhara, S.
Biofuels : Potential and Challenges	Pandey, A.K.
Biofumigation and Solarization for Management of Soil-Borne Plant Pathogens	Reddy, P.P.
Biointensive Integrated Pest Management in Horticultural Ecosystems	Reddy, P.P.
Chickpea (Gram) Status and Cultivation Technology	Shekhawat, S.S.
Controversial Herbal Drugs of Ayurveda	Saroya, A.S.
Diseases of Horticultural Crops: Nematode Problems and their Management	Reddy, P.P.
Flavouring and Fragrant Resources of India	Sood, S.K.
Flowers and Vegetables of India	Sud, R.K.
Folk Herbal Medicine and Drug Discovery	Katewa, S.S.
Fungal Diseases and their Management in Horticultural Crops	Reddy, P.P.
Hort-Agro Project Reports	Sud, R.K.
Horticulture and Livelihood Security	Nath, P.
Insect, Mite and Vertebrate Pests and their Management in Horticultural Crops	Reddy, P.P.
Integrated Nematode Management in Horticultural Crops	Reddy, P.P.
Ornamental Plants for Gardening	Chopra, V.L.
Plant Protection in Horticulture Vol. 1-3 (Set)	Reddy, P.P.
Productivity Enhancing Technologies for Horticultural Crops	Reddy, P.P.
Recent Trends in Disease Management of Fruits and Seeds	Prasad, M.M.
Salinity Tolerance in Plants: Methods, Mechanisms and Management	Garg, B.K.
Science of Tea Technology	Ahuja, P.S.
Silvipasture in India: Present Perspectives and Challenges Ahead	Palsaniya, D.R.
Tropical Forest Ecosystems Structure and Function	Singh, V.P.
Tropical Planting and Gardening 6th Edition	Barlow, H.S.
Useful Herbs of Planet Earth	Daniel, M.
Vegetable Gums and Resins	Howes, F.N.
Weed Seed Biology	Vanangamudi, K.

HORTICULTURAL ENGINEERING TECHNOLOGY FIELD MACHINERY

R.C. BALLS
ANCAE
National Specialist Mechanisation Adviser with the Agricultural Development and Advisory Service (ADAS) of the Ministry of Agriculture, Fisheries and Food, U.K.

United Book Prints

First Reprint in India

United Book Prints
4806/24, Ansari Road, Daryaganj
New Delhi - 110 002
Tel.: +91-11-41511055
E-mail: unitedbookprints@gmail.com

ISBN: 978-93-83692-01-9
Reprint : 2014

Reprint in India

CONTENTS

PREFACE

Mechanisation is becoming an ever more important aspect of horticulture, but after several years of giving engineering advice to the horticultural industry, I am aware that this is one of the least well understood of the many branches of science and technology of which the modern horticulturalist must have a working knowledge. The object of this book is to provide this working knowledge to both those in training, and those already established in the industry.

I have covered the subject from the basis of the application of engineering technology to the industry, using pure engineering only where it is in context. The aim is to provide the reader with broad principles, rather than with details of individual machines or basic engineering theory. Machine operating details can be obtained from manufacturers' handbooks or publications from trade associations and professional bodies; the basic engineering can be obtained from those textbooks on the specific subject.

The book is written in SI (Systéme International) metric units. Conversions to Imperial and other common units are given in appendix A.

Houghton Conquest
Bedford

R. C. BALLS

ACKNOWLEDGEMENTS

I would like to acknowledge the help given to me during the preparation of this book by colleagues in the Advisory Services and Research Stations. In addition I wish to thank the following equipment manufacturers for permission to use information contained in their leaflets: Ransomes, Sims and Jeffries (figures 2.7 and 2.8); Hestair Farm Equipment (figure 3.1); Lurmark Ltd (figure 4.4); Drake and Fletcher Ltd (figure 4.6); Union Carbide – Horstine Farmery division (figure 4.7); Grundfoss Ltd (figure 5.3); John Wilder (Engineering) (figure 6.1); Smallford Planters Ltd (figure 6.7); Eurotec Precision Ltd (figure 6.8).

Full use has been made of experience gained during advisory work on horticultural holdings in England, Wales and Scotland. I am indebted to my wife for typing the manuscript.

1 BASIC POWER UNITS

1.1 ENGINES

Most of the power applied to horticultural operations is derived from oil-based fuels by means of an internal combustion engine. Two basic types exist, spark ignition and compression ignition.

1.1.1 Types of engine

Spark ignition. The principle of these engines is that the charge of fuel is ignited by a spark from an external source. Since they are usually small, single-cylinder units fuelled by petrol, such engines are cheap to produce and maintain, and can be extremely low in weight for the power that they produce. These features make them attractive for light-weight domestic equipment and hand-held power tools. However, petrol is a relatively expensive fuel as there are no customs duty concessions. This type of engine can prove less reliable and require more frequent servicing than the compression type, but can tolerate greater variability in fuel quality. Routine servicing and repair is possible without complex specialist equipment.

Although petrol is the main fuel source, liquefied petroleum gas (LPG) can be used. This is a cleaner burning fuel, and its use is preferable in areas where people are working. Prior to the 1960s large, multi-cylinder, spark ignition engines were the basic power source of farm machinery; these ran on tractor vaporising oil (TVO). The virtual disappearance of this type of engine has led to the cessation of TVO refining.

Another fuel with enormous potential is alcohol distilled from material derived from plants. It has the advantage of being a totally renewable energy source, but at present is more expensive than oil-based fuel in temperate climates.

Compression ignition. In these engines the fuel charge is ignited by the heat produced when it is compressed in the cylinder; they are more commonly known as diesel engines. Most of the multi-cylinder and larger single cylinder engines used in European horticulture are of this type. They are much better suited to constant power demands and adverse environments than spark ignition types, and are more robust, but they are also more expensive to produce and are heavier than their

petrol counterparts. The fuel for this type of engine has to be free from water and dirt; also, major servicing demands specialist equipment.

Under U.K. fuel duty legislation, diesel fuel for 'off road' purposes is normally allowed concession from the taxes on road transport diesel (DERV); this makes it much cheaper than petrol.

Compression ignition engines can be run on a variety of combustible oils, including those obtained from rape and sunflower seed, but at present this has not been adopted commercially.

Turbo-charging. This is a technique for increasing the power of an engine. It is achieved by forcing a greater fuel/air charge into the combustion chamber by pumping air into the engine instead of allowing it to be drawn in by the induction stroke. The energy needed to drive the compressor is taken from the exhaust gases as they flow from the manifold. The effect of turbo-charging is a 15–20 per cent increase in power and a better specific fuel consumption, but with increased operating temperature and stress on components. Some production engines can be obtained in a turbo-charged version, and 'bolt on' kits are available for existing engines. Before installing the latter it is wise to check that the engine components are capable of bearing the extra strain.

1.1.2 Power characteristics

Power measurements. The correct choice of an engine or prime mover is impossible unless output characteristics are considered. A choice based purely on 'maximum horsepower' will inevitably be a bad one. The three essential parameters are

(a) Power, measured in kilowatts (kW).
(b) Speed, measured in revolutions per minute (rpm).
(c) Torque, measured in Newton metres (Nm). This is the twisting that the crankshaft can produce.

The way in which the engine was tested also has a bearing on the interpretation of the figures. The main points to bear in mind are as follows.

(i) The time during which the power can be sustained can vary between a very high figure, which can be sustained for only a few minutes (intermittent rating), and a more conservative figure, which can be produced for 12 hours or more (*x* hour rating). Road transport vehicle engines can be intermittently rated, whereas an engine for an irrigation pump will need a 12 hour rating. A tractor engine will normally have a 2 hour rating.

(ii) The engine might be tested without normal ancillaries, such as the alternator, so that some of the power quoted will be lost when these are being driven; figures determined in this way are quoted 'Bare Engine'. If tested with the ancillaries fitted, figures are quoted 'Engine Equipped'. Sometimes, as with tractors, the power reading will be taken from a convenient output shaft, and is termed 'Belt' or 'PTO' power; this gives a true reflection of the useful engine power.

(iii) The initials 'DIN', 'BS' and 'SAE' refer to the testing method used. There will be slight differences between the power obtained by the various methods.

Analysing power figures. Most engines used in U.K. horticulture have their power characteristics represented in graphical form. A graph for a medium-sized tractor is figure 1.1.

The most important line on the power graph for an engine to be used in horticulture is the torque curve. The speed at which the torque is at its maximum (1600 rpm) is considerably below the maximum governed engine speed (2000 rpm). This is intentional since it is a positive advantage in the operation of field machinery.

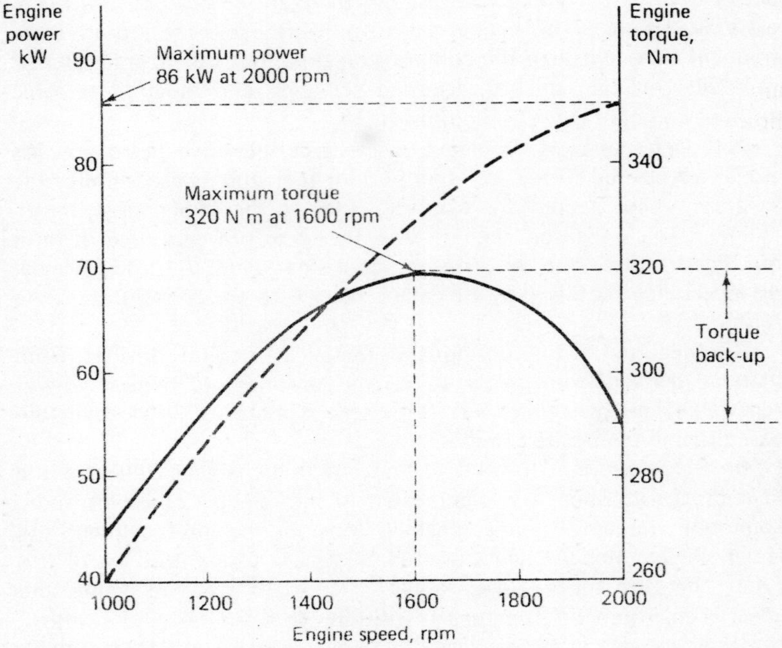

Figure 1.1 Typical engine power graph.

The torque output of the crankshaft is used to overcome the resistance of the machine that it is operating. Heavily loaded machines, such as rotary cultivators and mowers, are normally driven at near maximum engine speed, set to take full advantage of the engine power. If the machine encounters a slightly harder piece of ground or grass it will cause the engine to slow down and the crankshaft torque will increase, thus enabling the machine to 'slog through'. If the maximum torque coincided with maximum engine rpm, on slowing down the torque would also decrease and the engine would stall. The degree to which torque rises from that at the designated engine operating speed is termed the torque back-up. This can sometimes be found in manufacturer's data or official test reports (see table 1.1).

Table 1.1

Comparative data for three medium-sized tractors

	Tractor model		
	A	*B*	*C*
Construction			
Tyre size, front	7.50–16	7.50–16	7.50–16
Tyre size, rear	16.9/14–30	16.9/14–30	16.9/14–30
Unballasted weight, kg	3050	2271	2723
Engine			
Number of cylinders	4	4	4
Displacement, cm^3	3771	3863	3863
Rated speed, rev/min	2200	2000	2000
P.T.O. performance			
Maximum power, kW	41.5	46.7	42.1
At speed of, rev/min	655	652	669
Power at 540 rev/min, kW	39.5	42.2	38.3
Power at 1000 rev/min, kW	—	—	—
Torque back-up ratio, %	24.3	16.9	19.0
Fuel consumption			
At maximum power, l/h	14.28	13.4	13.17
Specific at max. power, g/kWh	292	240	262
Drawbar performance			
Maximum power, kW	35.5	39.8	38.7
Max. pull, ballasted or unballasted, kN	39.5	39.2	38.7
Hydraulic performance			
Maximum hydraulic power, kW	7.3	5.4	4.7
Flow at max. power, l/min	31.3	18.2	15.5
Pressure at max. power, bar	140	178	180
Lift capacity at implement	15.0	11.1	12.9

In a properly designed tractor or engine powered machine, the designated PTO speed or operating rpm should be above the speed of maximum torque. In some designs this is not the case; there are instances where a 50 kW tractor with a good torque back-up can outperform a 75 kW tractor with poor torque characteristics.

Table 1.1 shows the data needed to compare three engines as well as the tractors that they are fitted in. If only the 'maximum power' were being compared, one would rank them in the order B, C, A; however, for tasks that need 'lugging through' with a heavy PTO load they would be ranked in the order A, C, B.

1.1.3 Applications

There are three basic ways in which an engine can be utilised.

(a) Stationary. For such uses as driving irrigation pumps and electric generators, the engine will often be fixed to the floor or to a common baseplate with the equipment that it powers.

In these applications the loading is often at a constantly high level, so the power output must be chosen from the long hours duty rating. The constant load, and often indoor siting, mean that the cooling system must be more powerful than that for tractor or automotive duties.

To reduce the load when starting and running up to speed, a hand-operated clutch can be fitted. As the equipment is normally driven at one set speed, a 'constant speed' fuel pump or carburettor can be fitted. Because the engine is left unattended for long periods, a device can be fitted to stop it if the temperature rises or the oil pressure falls.

(b) For powering equipment. This applies to anything from a domestic lawnmower or chainsaw to an engine in excess of 150 kW on a pea viner.

In the case of domestic machinery or a chainsaw, the engine will normally be a light-weight petrol one; since these are produced in quantity, they can be designed to suit any particular piece of equipment. Good examples are the vertical crankshaft units for rotary mowers, and the very small two-stroke units with special carburation which enable a chainsaw to be used at any angle. Some engines have built-in gearboxes which provide a second power take-off to enable the machine to be driven forwards.

The large diesel engines fitted to field machinery are mostly installed with their crankshafts horizontal; a power transmission system is used to cope with any anomalies in driving positions.

Except where the machine can be taken out of work to ease starting loads, a clutch is essential to break the drive train. On small petrol engines this is accomplished by a centrifugal clutch or by allowing the drive belt to become slack.

(c) In tractors and vehicles. Here the engine is governed for automotive purposes, and is normally directly coupled to the transmission.

1.2 TRACTORS

1.2.1 Traction

1.2.1.1 Tractor configuration

The tractor should be able to transmit as much as possible of its engine power to the ground for tasks where only traction is required, such as ploughing or cultivating. There are three basic configurations.

Two-wheeled drive
Four-wheeled drive
Tracklayers.

In two-wheeled drive the tractor has a single driving axle, normally at the rear end, with the front steering axle undriven. The rear axle, rather than the front, is driven so that the weight and downward forces of implements affixed to the rear can add to its effective weight and improve traction. Most tractors have facilities to lock the rear axle differential so that both rear wheels can contribute to traction.

In four-wheeled drive configuration all wheels are driven. This improves the traction in two ways – all the tractor's weight is on driven wheels and there is no traction effort consumed in pushing undriven wheels through soft soil. The four-wheeled drive is characterised by three features: a driven steerable front axle, two non-steerable axle units joined with a vertical pivot to permit steering, and two steerable axles. The steerable front axle can be driven by shafts or hydraulically. The hydraulic system limits the amount of power that can be transmitted to the front wheels. On some shaft-driven units the drive is connected only in certain forward gears; also some arrangements cannot cancel the front differential drive action and are effectively 'three-wheeled drive.'

Tracklayers rely on a continuous track, which forms a 'roadway' for the tractor to drive on. The tracklayer puts an even horizontal force on a large area of soil instead of just localised areas under the wheels. This reduces the tendency for the soil to shear and for the track to lose traction. There is also less compression damage to the soil over which the machine passes.

To be fully effective the weight and draught forces of any attached implement should not transfer weight off the front end of the tracks.

To this end most tracklayers have hydraulic lift arms with a special geometry, or mounted implements are carried on a trailed carriage.

Some attempts have been made to add tracks to the rear wheels of conventional tractors. These have shown increases in performance, but have never had a commercial success.

1.2.1.2 Tyres for traction

It has long been accepted that increasing the weight on a driving wheel increases its tractive effect. Recent work by the National Institute of Agricultural Engineering (N.I.A.E.) at Silsoe, Bedfordshire, has quantified this tractive effect, together with the interaction of the inflation pressure and size of the driving tyres. An example of the results for two sizes of tyre is shown in table 1.2.

This work showed that a properly ballasted two-wheeled drive tractor could equal the performance of a similarly powered four-wheeled drive model with the standard ballast. This is important to market gardeners who cannot justify the purchase of a heavy-duty tractor for occasional tasks such as subsoiling but who might be able to use a small or medium tractor with the correct tyres and ballast. As well as traction efficiency, a correctly ballasted tyre will produce less wheel slip. In many cultivation operations wheel slip causes smear and soil panning.

The method for assessing the ballast for a two-wheeled tractor is

(a) Multiply the engine power (kW) by 100 to give the rear axle weight required in kg.
(b) Assess the weight on the rear axle with the pertinent implement attached and in use.
(c) Subtract (b) from (a) to get the required ballast to be added.
(d) Consult the rear tyre loading tables to see if these tyres can safely carry the load.
(e) If not, select a suitable size, ply rating and air pressure that will ensure safety.

Note that for very large tractors it is necessary to consult the manufacturer's maximum axle load ratings, to determine if the axles will carry the weight calculated in step (a).

Ballast can be added as water in the tyres or as bolt-on wheel weights.

For four-wheel drive tractors, divide the total weight requirement between the two axles, taking due note of the permissible loadings on the steering axle.

Example. Consider tractor A from table 1.1, fitted with a subsoiler.
(a) Quoted power = 41.5 kW; required rear axle weight is 41.5 × 100 = 4150 kg (2075 kg/wheel).

Table 1.2

Influence of tyre size, load and inflation pressure on tractive effect

Field conditions	*Load (kg)*	*Inflation pressure (kPa)*	*Pull at max. efficiency (kN)*	*Slip at max. efficiency (%)*
16.9/14–30 Tractor driving wheel tyre 6 ply				
Good	1750	83	7.0	10
Average			6.5	13
Poor			6.1	14
Bad			5.4	18
Good	2000	103	7.9	10
Average			7.3	13
Poor			6.8	15
Bad			5.9	19
Good	2280	131	9.0	10
Average			8.2	14
Poor			7.6	16
Bad			6.4	21
16.9/14–34 Tractor driving wheel tyre 6 ply				
Good	1850	83	7.4	10
Average			6.8	12
Poor			6.5	14
Bad			5.8	18
Good	2000	90	8.0	10
Average			7.3	13
Poor			6.9	15
Bad			6.1	19
Good	2200	110	8.7	10
Average			8.0	13
Poor			7.4	15
Bad			6.5	20
Good	2420	131	9.6	10
Average			8.7	13
Poor			8.0	16
Bad			6.8	21

(b) Rear axle weight of tractor = 2000 kg; weight of subsoiler = 500 kg; assumed draught weight transfer = 500 kg; total = 3000 kg.

(c) 4150 minus 3000 = 1150 kg ballast requirement.

(d) From the tyre data tables a 16.9/14–30 tyre will carry 2000 kg at 103 kPa pressure.

Thus the maximum tractive performance of tractor A when subsoiling will be obtained by adding 1150 kg ballast and running the tyres at 103 kPa pressure. There is no need to change the tyres supplied.

Soil type can influence the choice of a tyre for traction purposes. Loose sandy soils require a wider, low-pressure tyre with shorter lugs; this prevents wheel slippage, the occurrence of which causes the tractor to dig itself into the ground.

1.2.2 Directly driven equipment

1.2.2.1 Mechanical power take-off (PTO)

On most modern tractors this is in the form of a rotating shaft at the rear of the tractor; belt pulleys are now very seldom fitted. Some tractors also have an additional forwards facing shaft to facilitate the drive for front-mounted machines.

On small and medium tractors the PTO shaft has 6 large splines and is designed to rotate at 540 rpm. This design of shaft proved incapable of transmitting power much over 50 kW, so a higher-power version evolved that has 21 small splines and runs at 1000 rpm. Some medium tractors have the facilities for fitting either a 540 or a 1000 rpm shaft. Care must be taken to avoid running 540 rpm based equipment at 1000 rpm, since serious damage will occur.

A PTO shaft can be driven directly from the engine or from the rear axle final drive. The latter option gives the facility of driving a machine at a speed proportional to its forward speed.

Some tractors have the PTO drive through a two-stage clutch, whereby the initial pedal depression stops the forward motion, and full depression stops the PTO. This is often termed 'live PTO'. On other tractors the PTO clutch is operated by a lever, totally independent of the forward motion clutch pedal. In both cases the tractor can be stopped, or put into a different gear, without stopping the machine that it is driving. This is desirable when using machines that need to run until empty before they are stopped.

1.2.2.2 Hydraulic power

Tractors with hydraulic lift linkage have the ability to power small hydraulic motors. The power from an 'in-board' pump is normally limited to less than 10 kW; higher power can be obtained by adding external pumps on the PTO or the engine crankshaft. This method of power transmission is ideally suited to driving remote items or those that have to move around in the course of operation.

Hydraulic drives can offer infinite speed control in either direction but, if more than one motor is driven from a single pump, the motor speeds will be uncontrollably varied in relation to the loading on each motor. When compared with hydraulic systems used in other industries tractors operate at very high pressures, between 15 and 20 MegaPascals (MPa) or 2100 and 2900 psi. For this reason, hydraulic components from other industries should only be used on tractors after careful thought has been given to their suitability.

1.2.2.3 Electric power

Very limited power can be taken from the normal electrical system. This is usually supplied at 12 V d.c.; most d.c. motors have speed characteristics related to load, therefore this system is unsuitable where accurate speed is important.

When mains-driven fixed equipment is used in the field, its power can be derived from a PTO mounted generator.

1.2.3 Gearing and power transmission

1.2.3.1 Gear ratios

The varied gear ratios available on tractors and other self-propelled machines are necessary for three reasons.

(a) To maximise efficiency. As explained earlier, the maximum power output occurs over a relatively small part of the running speed of an engine. On the tractor output graph of figure 1.1, the engine will need to run between 1600 rpm and 2200 rpm to be in the correct part of the torque curve. The tractor could probably plough at 5 km/h, but subsoil at only 2 km/h. In both cases the best gear ratio should be chosen to allow the engine to run at 2000–2200 rpm.

(b) To vary speed. A tractor will need to be able to travel from a high speed for road transport to a crawling speed suitable for hand transplanting. This degree of speed variation is beyond the flexibility of the engine alone, which will probably be unable to produce a variation of greater than 1:6.

(c) To match machine speed with forward speed. Many trailed PTO driven machines have to work at a set speed. To cope with variations in loadings and operation the forward speed will have to vary; this is accomplished by the gearbox. On some machines the forward speed/drive speed ratio will affect the job being carried out. A good example of this is a rotary cultivator where a slow forward speed/high rotor speed will produce a fine tilth at a slow rate. A high forward speed will produce a cloddier tilth at a higher rate.

1.2.3.2 Types of speed ratio change system

(a) Drive belt pulley. On simple machines such as small rotary cultivators, speed ratios can be altered by moving the drive belt between pairs of different sized pulleys. This system requires the machine to be stopped and components to be detached for ratio changing. It is possible to have a 'change on the move' belt speed variator – split vee pulleys used in conjunction with a wide vee belt. The degree of split of the pulley decides how far into the groove the belt lies, and thus its effective diameter (figure 1.2). These systems require accurate sliding components which can easily seize with rust if used on outdoor equipment.

(b) Gear or chain wheel changes. These are used in similar fashion to (i), except that there are no methods for 'on the move' changing.

(c) Sliding spur gearbox. This is the traditional gearchange. In older systems the gears were moved into mesh to change speed; in the syncromesh box the gears are constantly enmeshed and connected to the

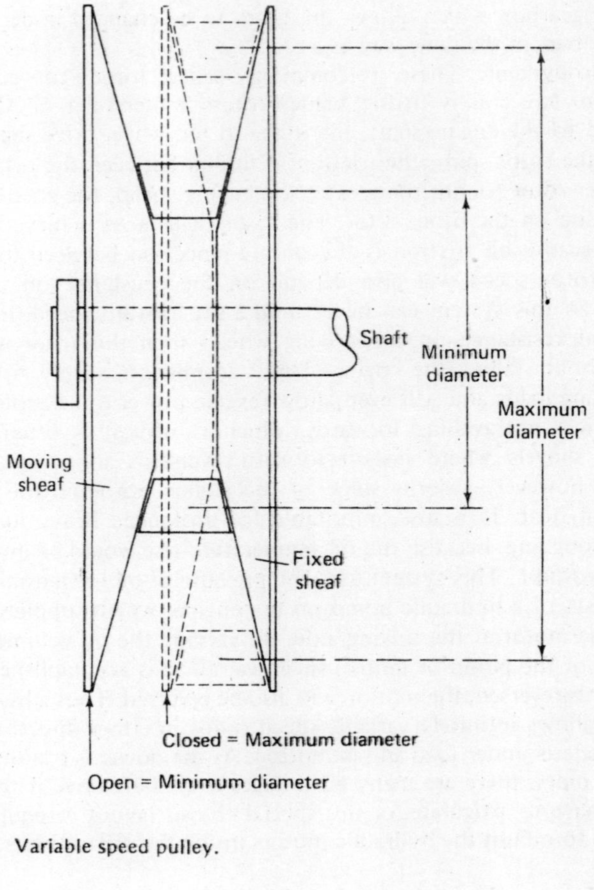

Figure 1.2 Variable speed pulley.

shaft by sliding dog clutches. The latter system suffers less damage if the gears are changed under load, as the ensuing 'crash' will not cause the gear teeth to be chipped. Many tractors have in excess of 10 speed ratios. These are not arranged in a single 'box', but consist of two or three interlinked boxes; thus a 12 speed box might be a 3 speed x 2 speed x 2 speed, each with its own lever. One of the 2 speed boxes might also contain the reverse gear.

(d) 'Change on the move'. Some tractors have a main sliding spur box with one of the two speed boxes having the ability to change under load and motion. This enables a tractor to be changed down without loss of motion if the field conditions worsen, the higher speed being used where conditions are lighter. Most of these systems use constantly meshed planatory (epicyclic) gears where their connection to the output is by means of plate clutches. Some tractors are fitted with a complete 'change on the move' gearbox which is an extension of the system described above. A further development is the fully synchromeshed gearbox which allows the gears to be changed under motion, but not load, in the same way as a car.

(e) Hydrodynamic. These are commonly called 'torque converters' and consist of two closely fitting, bladed rotors immersed in oil. One rotor is fitted to the engine shaft, the other to the wheel drive shaft. When the engine rotor spins, the friction in the oil between the rotors drags the other rotor round; the greater the engine speed, the greater will be the torque on the other rotor. The rotors will never achieve the same speed because oil friction is the only connection between them. The driven rotor speed will also depend on the resistance on the drive wheels, so this system can never hold a set forward speed. If there is sufficient resistance on the driving wheels then this rotor will stop, but without stalling the engine. This transmission is very tolerant of fluctuating loads and will even allow reverse power to be applied while the vehicle is travelling forwards, which is why it is often used in loading shovels where instant forward/reverse is an advantage. The system, however, absorbs some of the engine power in the form of frictional heat. It is also unsuitable for prolonged heavy loading, as when ploughing, because the oil temperature rise would be intolerable.

(f) Hydrostatic. This system uses the pressure of oil to transmit power. It consists of a hydraulic pump on the engine, which supplies oil to a hydraulic motor at the driving axle. By varying the oil volume characteristics of the pump or motor, speed variation is accomplished; if the oil flow is reversed, the motion will also be reversed. Thus a hydrostatic system allows infinitely variable speed ratios in either direction, while the vehicle is under load and in motion. As the power is transmitted by flexible pipes, there are many advantages over shaft drive if the driven wheels have to articulate, or if a special chassis layout is required. It is possible to mount the hydraulic motors inside the drive wheels to avoid any shafts. If a pair of hydraulic motor wheels are mounted on an axle with castor steering on the other, the vehicle can be steered by varying the flow between motors and thus the wheel speeds.

Hydrostatic power with variable flow operates at a fixed oil pressure. In most systems the motor torque is related to oil pressure, so there is no increase in torque resulting from 'changing down' as would be expected in a gearbox. Hydrostatic transmissions are more expensive than conventional gearboxes, because of the precision parts involved. Some power is also lost as a result of oil friction and internal leakage.

1.2.4 Avoiding damage

In addition to heavy draught work, tractors are called on to operate in soft tilths without excessive sinkage and soil compaction damage.

1.2.4.1 Beds

In field vegetable work a system has been evolved whereby no tractor wheels pass over soil in which the crop is grown. The field is divided into 'beds' where the crop is grown, and 'wheelways' where tractor and heavy implement wheels run. As the land lost to wheelways should be as little as possible, narrow tractor wheel are preferable. These will cause severe compaction as the tractor weight is concentrated on a small area, but as this happens in the wheeling it will not damage areas where the crop is grown. Rear tyres can be obtained that are only 200 mm or 250 mm wide, and these should be used if their load-carrying capacity is sufficient.

Under certain conditions a very narrow wheelway will be difficult to follow because the tractor will tend to run off into the softer ground on either side. If this is a common occurrence, wider basic wheelways of 300 mm or 350 mm should be formed.

1.2.4.2 Low ground pressure systems

This method is of interest to horticulturists who do not use the bed system; the tractor wheels run over areas where the crops will grow, and any compaction caused will be detrimental. Several methods have been used to minimise soil compaction.

(a) Tracks. As stated previously, a good tracklaying tractor will spread its weight evenly over the entire ground contact area of the tracks. Several small, light-weight tracklayers are available; the essential feature for low pressure is the maximum number of small idler rollers along the base of the track which help to spread the tractor weight on to the track base. Some operators have significantly increased the base area of the track by bolting lengths of timber across the track plates. The main problem with a tracklayer in this context occurs when it turns. The tracks have to slide sideways during turning, which can cause bulldozing of the soil into curved ruts. This feature makes a tracklayer unsuitable for travelling on soft turf, for example to top dress it in early spring, as it would badly cut up the turning areas.

(b) Tyres. Recently much interest has been shown in wide profile tyres running at low pressure. Tractor rear tyres in excess of 1 m wide can be obtained and, if run at a suitably low pressure, show less sinkage than a human foot on the same soil. Although soil pressure is difficult to measure, most experts agree that it is unlikely to be less than the air pressure in the tyre. It is, at present, considered that a tyre pressure of 28 Pa should be the maximum. At very low pressures there will be little binding force between the tyre and the wheel rim, so under heavy draught conditions the rim can turn within the tyre and rip out tube valves.

It is considered that the small degree of compaction caused by the first wheel to travel over the soil will aid its resistance to compaction from subsequent wheels, so a two-wheeled drive tractor should ideally have front tyres as wide as those at the rear.

(c) Special wheels and attachments. The standard form of reducing soil compaction is the 'cage wheel'. This is used to double the width of the rear tyres of a tractor. The cage wheel has an outer rim of steel bars (hence the term 'cage') and normally has some form of rapid coupling. To avoid the steel rim being damaged by hard roads, it is normally 20 mm less than the running diameter of the normal tyre. This means that the normal tractor wheel has to sink by 10 mm before the cage carries any weight.

Secondary wheels to match the existing tractor rear wheels have become available recently, with the same rapid couplings as cage wheels. These do not have the same pre-sinkage problems as the cage.

Recent research into soil compaction has resulted in new designs of cage wheels which replace the normal rear wheels rather than supplement them, so the tractor is unable to travel on the highway. Various cage bar shapes and orientations were tried, the most successful being thin radial 'paddles', but these offered no support against sinkage, so the production wheels are a compromise between this design and a conventional cage bar for support.

1.2.4.3 Growing crops

Where tractors have to travel through growing rows of crops, two methods are available to minimise damage. The first is to use the narrow wheels mentioned above. Where the tractor weight is too great to be supported by a pair of narrow wheels, it is possible to twin a second narrow wheel on to the rear axle with a spacer, so that the second wheel runs in the space between the next pair of rows. Where tall crops have to be traversed, high clearance adaptations are available. These might consist of over-sized rear wheels or downwards projecting final drive units on the rear axle, used in conjunction with a high clearance front axle.

In orchards and vineyards, special narrow versions of tractors are available to allow passage between plant rows.

1.2.4.4 Turf

Damage to turf can occur from cutting up, as well as wheel sinkage, so most tractors used on turf have wide tyres with smooth treads. A smooth tread is allowable as most tasks will not require high draught. New agricultural rear tyres should be avoided as the lugs will dig out divits of turf.

1.2.5 Hydraulic take-off

Most tractors have facilities for tapping the hydraulic system output for powering rams or motors, and are fitted with a 'spool valve' assembly for this purpose; the facilities provided on this valve will affect the way in which the external hydraulics can be used (see table 1.3).

Table 1.3

Spool valve options

Type of valve	*Affect on rams*	*Affect on motors*
Single acting	Power one way, return by gravity/spring. Can stop midway	Drive in one direction. Stop/ start only but will prevent motor running backwards
Double acting	Power both ways. Hold at any position	Drive either direction. Stop and hold from either direction
Double acting with float position	As for double acting, but float allows ram to move freely	As for double acting, but motor will idle if necessary
Flow control	Speed of action variable	Speed variable

1.2.6 Safety

The main safety feature on tractors used in the U.K. is to protect the driver against crushing should the machine overturn. On most machines this involves a clad cab also fitted with acoustic insulation to limit the driver's exposure to noise. On tractors that have to operate in confined positions, such as glasshouses, orchards or vineyards, simple rollbar frames are permissible. If headroom is limited it is also permissible to hinge down the top part of the bar, but this must be reinstated as soon as the tractor returns to open land. Some farm vehicles are exempt from safety frames; these can include forklift trucks and tracklayers.

The safety frame forms an integral part of the tractor, so only frames or cabs that have been proved satisfactory when attached to a particular make and model of tractor are recognised as satisfactory.

Other major safety features involve the guarding of rotating parts, especially the PTO, so that limbs or loose clothing cannot become entangled.

Full details of the current U.K. safety legislations can be found in publications of the Health and Safety Executive (see appendix B).

1.3 ELECTRICITY

The main power source for equipment used in and around buildings on the nursery is electricity. Additionally it is the major source of power for artificial illumination and some heat production. It is usually obtained from the public supply grid, although in some circumstances it might be produced on the premises by generator. The near monopoly of public grid supply means that all commonly available electrical equipment is designed to match the grid supply. Most independent electric generator sets also supply power to match the grid supply in all vital aspects.

1.3.1 Electricity supply

1.3.1.1 Basic characteristics of supply

Public mains supply operates, nowadays, on the alternating current (a.c.) system, rather than the direct current (d.c.) system. Direct current power is of the type that is supplied by a battery; here the voltage between the two supply wires is constant and always of the same 'polarity'.

In the a.c. system, the voltage in one conductor alternates in polarity with respect to the other conductor; this is depicted in figure 1.3. In the U.K. and most other European countries, the a.c. mains supply alternates with a frequency of 50 cycles per second (50 Hertz, Hz); in the U.S.A. the mains frequency is 60 Hz.

One conductor of the supply transformer is connected to the ground; this is the 'neutral' wire of the supply and is regarded as having a voltage potential of 0 V. The voltage in the other conductors, termed 'line', 'live' or 'positive', varies between +340 V and −340 V with respect to neutral in European mains supplies. These are, however, peak values; the resulting mean voltage in the line wire will be (peak voltage)/$\sqrt{2}$ or 240 V, which is often termed the 'root mean square' or r.m.s. voltage. In the U.S.A. the supply to many small nurseries is 110 V a.c.

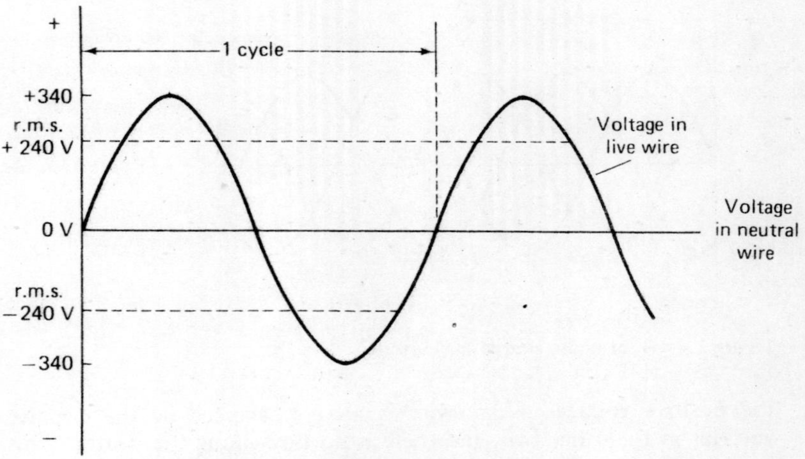

Figure 1.3 Single-phase alternating current.

The a.c. supply can have a single line output (single phase) or a number of line outputs superimposed (polyphase) on each other. The most common polyphase is three phase, although occasionally a supply might be termed 'two phase'.

1.3.1.2 Single phase

This, with the wave form shown in figure 1.3, is the normal supply to 'domestic' and small industrial premises, which will encompass many small nurseries. The main problem with a single-phase supply is that its characteristics for operating motors are poor compared with three phase, and its supply size is limited. In the U.K. the maximum supply available is often 50–60 amps, allowing only 12–14.4 kW to be drawn. This, combined with the poor starting characteristics of single-phase motors, dictates a maximum motor size of around 1.5 kW.

1.3.1.3 Three phase

To reduce the cyclic pulsing of a single phase a.c. supply, the main generators have three pairs of 'poles' which superimpose three single-phase sine waves within one cycle (figure 1.4.).

The main difference between the three-phase and single-phase graphs is the available voltage. On the single-phase graph the voltage between the wires will rise to 340 V and then decrease to 0 again every 1/100th of a second. If one analyses the three-phase curves, the peak voltage between any pair of 'coloured' wires will be 600 V, giving a corresponding r.m.s. voltage of 415 V. However, the voltage between any wire and earth (neutral) is still 240 V. One further characteristic is that

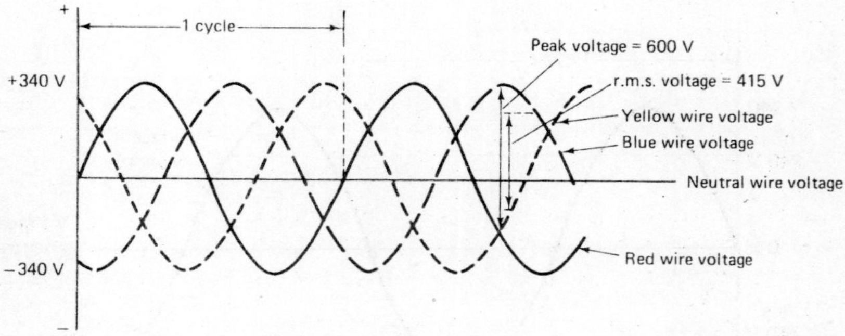

Figure 1.4 Three-phase alternating current.

the positive voltage in one wire is always balanced by the negative voltages in the other two, and there is no flow along the neutral. This is why a three-phase requires only three single wires, and not three pairs of 'line' and'neutral'.

As the power flow to a motor is even, it can start more easily under load, and its physical size (and hence cost) is much smaller for a given power than a single-phase motor.

Where possible, all three-phase installations should be arranged for even distribution between phases. Thus a 9 kW heater will be divided into 3 x 3 kW banks, each running from one phase. As stated above, a three-phase supply to a building or machine will need only three conductors, although this will supply only 415 V. Should single phase also be needed, a four conductor supply is necessary, the fourth being a neutral conductor. Although it is possible in a three wire supply to transform back from 415 V to 240 V, the 'neutral' obtained will not necessarily be at 0 V; and a three pin, single-phase plug wired from it could be highly dangerous, with its 'neutral' pin being 'live'. Both three and four wire supplies must also contain the earth continuity wire, so they will be represented by a four pin or five pin plug/socket, respectively.

1.3.1.4 'Two phase'

In the two-phase supply, the centre of the transformer output winding is grounded; this produces two single phases of opposite polarity, with a peak voltage of 740 V (480 V r.m.s.) between the two line conductors. The voltage of either conductor with respect to ground is still, however, only 240 V r.m.s. In this supply the phases are peaking at the same point in the cycle, rather than at the staggered intervals of a true polyphase, thus the rotating field effect of three phase when applied to a motor is lost. It is therefore more correct to refer to this supply as being '480 V single phase'.

1.3.2 Power formulae

The relationship between watts (W), amps (A) and volts (V) is

watts = amps × volts

For example a 240 V supply rated at 10 A will be suited to run a heater of 240 × 10 = 2400 W or 2.4 kW.

In a.c. supplies the power can also be derived by multiplying the r.m.s. values of voltage and current, but this is only correct if the cycles of voltage and current occur simultaneously. In most a.c. machines the cyclic alternation of current induces an opposing voltage (back e.m.f.) which is out of phase with the applied voltage; this has the effect of altering the phasing of the current with respect to the applied voltage. This is shown in figure 1.5 and it can be seen that, for part of the cycle, the voltage and current are in opposition; hence no power is being produced and the available power will be less than the r.m.s. derived power. The ratio between the two power values is known as the 'power factor' (p.f.).

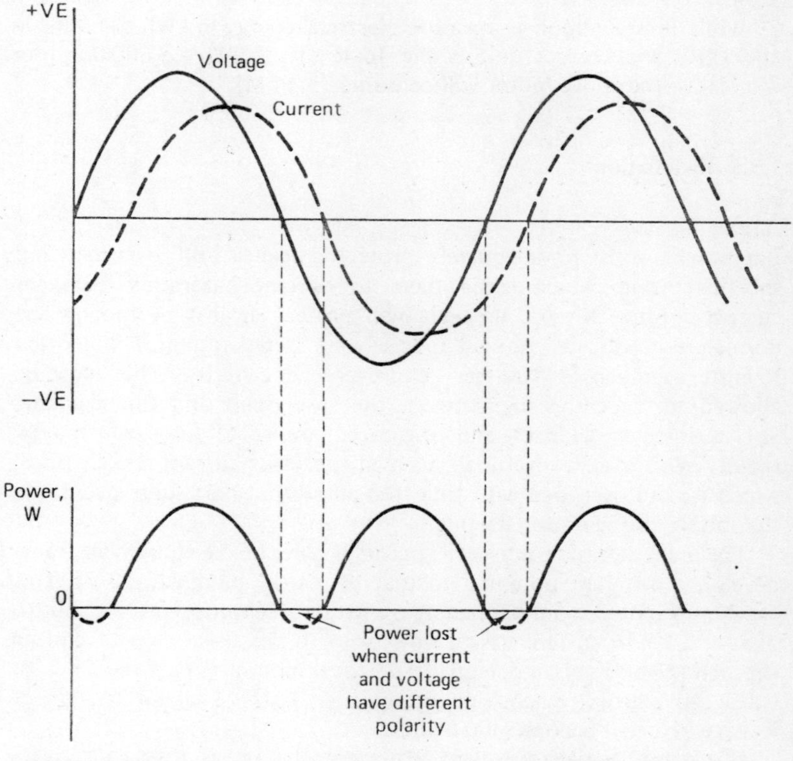

Figure 1.5 Effect of phase shift (power factor) on power in an a.c. circuit.

It is normal to identify the r.m.s. derived value by the unit 'volt amps' (VA), therefore

$$p.f. = \frac{W}{VA}$$

For example, a single-phase a.c. motor with a p.f. of 0.85 drawing 10 A at 240 V has a useful power input of 10 x 240 x 0.85 = 2.04 kW.

The power factor should not be confused with 'efficiency', which is the ratio between power input and the useful power output. For example, if the above motor has an efficiency of 90 per cent, the useful output will be

$$2.04 \times \frac{90}{100} = 1.84 \text{ kW}$$

The energy consumed is measured in watt hours (Wh), which is calculated by multiplying the power consumed by the hours of operation. For example, a single-phase a.c. motor with a p.f. of 0.85 drawing 10 A 2.04 x 10 = 20.4 kWh.

While it is common to measure electrical energy in kWh or 'units' in the U.K., the correct unit is the Joule (J); 1 kWh = 3 600 000 J or 3.6 MJ, so the above motor will consume 73.44 MJ.

1.3.3 Installation

1.3.3.1 Fusing

Equipments must be adequately protected against both over-load and short-circuiting, which cause them to become hazardous. Fuses or current-sensitive circuit breakers will protect against over-load. The normal fuse will not blow at precisely its stated rating; it will often tolerate as much as 100 per cent over-lead capacity. This must be allowed for when wiring between the fuse point and the machine. Special fuses are available and range from 'anti-surge' fuses, which blow rapidly with a little over-load, to high rupturing current (HRC) fuses, which equate over-load with time and allow the short surge over-loads that occur during motor starting.

The main hazard to personnel occurs if parts of the equipment frame develop a different potential to that of 'earth' (nominally 0 V). This can be prevented from happening by fitting an appropriate fuse, but to blow it a heavy current has to flow to earth and this might take place through the body of an operator. The safest method is to instal an earth leakage or current balance circuit breaker, both of which can detect leakage before it becomes hazardous.

It was traditional to use any water pipe system as an 'earth return'. This is now no longer safe, as many pipes are made of non-conductive plastics. Instead, plumbing should be connected to the circuit protection (earthing) conductor to detect any electricity leakage into the water system.

1.3.3.2 Wiring

Electric cables, whatever their composition, have some resistance to the flow of current. If the resistance is excessive it will cause heating and there will be a drop in voltage, resulting in either the cable insulation melting or a motor failing owing to insufficient voltage.

In the U.K., regulations drawn up by the Institution of Electrical Engineers (IEE) lay down strict rules as to the sizing of cables. These take into account the rated current, length of cable and other factors such as how much the cables are thermally insulated.

Cable-covering materials can be specified to suit the intended environment, the commonest being the plastic material PVC. For high-temperature operations, other plastics have to be used, and for maximum protection against heat, physical damage and water ingress the mineral-insulated, copper-covered (MICC) type is used. Many fixed installations use conduit or trunking to carry cables neatly; this should be done with caution when they pass through buildings with different environmental conditions, as it is possible for water to condense inside the conduit and run into switches.

1.4 ELECTRIC MOTORS

1.4.1 Types of motor

Two basic motor types are used in horticulture, universal a.c./d.c. and induction.

1.4.1.1. Universal

These are able to work on either a.c. or d.c., but are commonly found only on small power tools. All electric motors rely on magnetic attraction between the centre rotating member (rotor) and the case (stator) creating the torque to rotate them, this magnetism being created electromagnetically. On the universal motor, electromagnetic coils in both the rotor and the starter are energised, the power to the rotor coils being supplied through a sliding contact on the shaft (the commutator). The commutator is divided into sections, each wired to a part of the rotor coil, so the power is applied to each of these coil sections as it is attracted to the magnetic pole created by the stator coil. As it rotates, the commutator sectioning changes the magnetism to the next coil section, so the rotor is always under rotational attraction.

This motor is suitable for a.c. or d.c. as either form of power will energise electromagnets. The speed of a universal motor depends on its load, and the power supplied to the electromagnets. Thus an electric drill or hedge trimmer will run at a fast speed out of work, and slow down when in work.

Most of these motors operate at a high speed (up to 30 000 rpm), with high ratio reduction gears to bring them to operating speed. Speed control is normally achieved by varying the power to the motor. Although these motors contain sliding contacts on to the commutator in the form of self-lubricating carbon brushes, they should not be regarded as capable of constant operation.

Universal motors can be obtained for 12 V or 24 V operation, for battery equipment or for 110 and 240 V mains motors.

1.4.1.2 Induction motors

Like the universal motor, these operate by magnetic attraction between stator and rotor; the difference is in the way that the rotor is energised.

The induction motor will work only on a.c. for two reasons. First, the stator coils are arranged so that the sine wave pulsing travels around it in a circular direction, causing rotating magnetic poles. Second, when these magnetic waves pass into the rotor, they 'induce' it to generate its own electricity which in turn creates magnetic poles within the rotor. In this type of motor, the poles in the stator move while those in the rotor are static.

The torque of an induction motor depends on the current induced into the rotor coils. This will depend on the difference between the speed of the rotor and the speed with which the magnetic poles are rotating in the stator. The greater the speed difference, the heavier the current will be. As there will be no induced current when the rotor speed synchronises with stator pole speed, the rotor will never, in practice, reach it.

The speed of the poles in the stator is related to the frequency (cycles per second) of the power supply, and the number of poles created by the coil winding pattern. With the U.K. frequency of 50 cycles per second, a rotational speed of a pair of poles is 3000 rpm. The speed difference needed to cause rotation will normally result in a rotor output speed of 2800–2900 rpm. Most induction motor output speeds are related to the number of pole windings, the common ones being

2 pole – synchronous speed 3000 rpm; practical speed 2800–2900 rpm
4 pole – synchronous speed 1500 rpm; practical speed 1400–1450 rpm
6 pole – synchronous speed 1000 rpm; practical speed 900– 960 rpm
8 pole – synchronous speed 750 rpm; practical speed 700– 720 rpm
10 pole – synchronous speed 600 rpm; practical speed 520– 550 rpm

These speeds apply to a 50 cycles/second supply; in the U.S.A. the same motors would run 20 per cent faster owing to their 60 cycles/second supply.

As the rotor current is proportional to speed differential, an induction motor cannot tolerate a large reduction in speed, as a result of over-loading, without inducing a current in the rotor that is too high for its coil windings.

Speed control of induction motors is a complex subject. Two or three speeds can be obtained by wiring the stator so that a varying number of pole pairs can be chosen. Voltage regulation can be used on small motors (less than 1 kW) but their torque will also fall, so applications must be carefully chosen. Other devices such as frequency changing and slip control are beyond the scope of this book.

1.4.2 Starting electric motors

When starting, a motor must provide sufficient torque to overcome the resistance and inertia of the component it is driving. The current required to produce this starting torque will be greatly in excess of the current absorbed when the motor is running at full speed. The degree of excess starting current will depend on the motor type and the starting system fitted. These characteristics will often determine the strength of a power supply, rather than the total running load.

1.4.2.1 Single phase

Single-phase induction motors need some means of initiating rotation, the most common being 'split phase', where a second set of 'starter' windings is incorporated into the stator. Starter windings are connected to the supply via a capacitor which sets their magnetic field out of phase with the main windings. This provides an unbalancing force which causes the rotor to turn. When the motor is at full speed, the starter windings impede its efficiency and so are automatically switched out by a centrifugal switch.

A single-phase motor is started by connecting its input wires to the mains supply, known as 'direct on line' starting. The main components of a starter are a heavy contactor, capable of handling the peak starting current, an over-load current sensor and a 'no volt' sensor, both of which release the contactor.

A split-phase capacitor start motor will consume five to six times its running current for starting. For example, a 0.5 kW motor will consume (500 W)/(240 V) = 2.08 A to run, but will take up to 2.08 x 6 = 12.48 A to start. Thus an over-load sensor or fuse must tolerate the starting peak. This makes most current trip devices unsuitable for detecting small running over-loads which could cause a motor eventually to burn out. A more reliable method for preventing motor burn-out is to use a thermally operated switch within the motor windings.

The direction in which a single-phase induction motor runs is determined by the relationship between the starting and running windings. Thus the motor will run in the same direction, whichever of the two input wires is connected to live. To reverse direction, the leads from the starting and running windings must be transposed within the motor terminal box. These leads are normally labelled A1 and A2 for the running winding ends, and Z1 and Z2 for the starter winding. If a motor runs clockwise with A1/Z1 and A2/Z2 connected together, it will run anti-clockwise with A1/Z2 and A2/Z1 connections.

1.4.2.2 Three-phase motor

As the windings within this motor create a rotating magnetic field, it will self-start. However, like the single-phase motor, its starting current will be higher than its running current.

If a three-phase motor is started 'direct on line', it will have a starting current 5 to 6 times greater than the running current. However, it is possible to reduce this by using a 'star–delta' starter, in which the windings of each phase are initially connected in sequent pairs across each phase (star), then singly (delta) when it has reached a certain speed. By using the star connection, the voltage across each winding is reduced to $1/\sqrt{3}$ or 0.58 of normal, which causes a corresponding reduction in the current consumed. The starting current of a star–delta system is 3 to 3.5 times its running current. The motor is unable to develop its full speed or power on star, and thus the windings are reconnected into delta pattern once the motor is running at 75–80 per cent of its full speed.

The star–delta transition can be mechanically or manually operated. The manual starter box has an operating lever with 'start' and 'run' positions which is held over into 'start' until the motor is heard to be running at a constant speed, then is pushed to 'run'. The lever is held in 'run' by a magnetic coil which will release it if the running current is too great; thus if the change from 'start' to 'run' is made too rapidly, the lever will not stay in the run position. The mechanical star–delta starter uses a timer which will hold in star for a certain number of seconds, this time being adjustable to suit the machinery being driven.

1.4.2.3 Starter accessories

In direct on line or mechanical star–delta starters, the contactors connecting the main current are operated by electromagnetic coils. These coils require little power to operate, and allow the motor to be started remotely without taking the heavy current cables to the control point, or allow the controller to operate at a safe, low voltage.

All motors, however small, should be operated through a starter with a 'no volt release'. This ensures that a motor that stops as a result of overload or supply failure cannot restart automatically on resumption of supply. Further safety can be ensured by interlocking the starter with safety switches on doors or guards so that a machine cannot operate when anyone is in a vulnerable position. A good example of this is found in bulk crop store ducts, where switching on the duct lights breaks the fan starter circuits.

Machine operators can be protected by safety stop buttons placed at suitable points on the machine or within the work location. This type of emergency stop should 'lock off'; that is, once pressed the supply cannot be restored until the button itself has been reset.

1.5 ELECTRIC GENERATORS

These are mainly used as a source of mains stand-by. They can either be driven from the tractor PTO, or permanently linked to an engine.

Most generators consist of a basic alternator plus the necessary overload protection, and meters for output voltage and frequency. The alternator normally runs at around either 1500 or 3000 rpm to provide the 50 cycles/second mains frequency. This allows direct drive from an engine; normally only small units run at 3000 rpm, and suit higher-speed petrol engines. PTO sets have to have a speed-increasing drive unit.

1.5.1 Calculating generator requirements

A generator must be carefully selected. In terms of power supply it is 'on its own', and has not the peak surge back-up of the mains supply. If sizing a stand-by generator, the following steps are useful

(a) Assess which services are essential.
(b) The base load in kW will be motors (taking 1 hp = 1 kW) plus heat, lights, etc. in kW.
(c) A large motor should be taken as kW = hp x 3 to take account of starting current.
(d) If the nursery contains one large motor and several small ones, plus light and power, the generator could be sized to start the large motor and then take on the rest of the load when it is running.

For example, suppose a nursery has a cold store with a 10 kW compressor, 4 x 1 kW fans, a 2.5 kW heated bench and 5 kW of nightbreak lighting, all of which must be protected against mains failure.

The loads will be either

starting the compressor = 3 x 10 kW = 30 kW

or

running all the loads = 10 kW (compressor)
4 kW (fans)
2.5 kW (bench)
5 kW (light)

21.5 kW

The nursery can be run by a 30 kW generator, provided that the compressor starts before anything else is switched on.

However, as the compressor will be on a thermostat, the practical load will be

4 kW (fans)
2.5 kW (bench)
5 kW (lighting)
30 kW (starting compressor)

41.5 kW

Thus a better unit would be a 41.5 kW generator.

If a large nursery with some essential services is to be protected, a means of isolating the non-essentials will have to be installed. This is particularly important if an auto-start mains failure unit is used.

Generator output is normally quoted in kVA, which differs (slightly) from kW by a 'power factor'. This depends on the type of load, but it is normal to assume a power factor of 0.8, therefore the above generator will be rated at 41.5/0.8 = 51.9 kVA, or practically 52 kVA.

1.5.2 Generator power requirements

The engine to drive a generator must produce at least 1.25 kW per kVA of capacity, to allow for losses in the transmission. Where a tractor is to be used for driving a generator over a prolonged period, its available power should be taken as only 70 per cent of its rated output to prevent excessive wear; this in effect derates its power output to that which the engine would be considered capable of if sold as an 'industrial unit'.

This means that the foregoing 52 kVA generator will need an industrial engine of at least 65 kW, a tractor with a 65 kW PTO rating for stand-by use, or one of 93 kW PTO rating if used as a prime power source.

In terms of tractor workload, the above duty is similar to pulling a 3 or 4 furrow plough, therefore a tractor in good condition should be used.

1.6 PORTABLE POWER TOOLS

1.6.1 Electric drive

One common method for powering tools, such as portable hedge trimmers, is to use electricity. This can be taken from the mains, with a trailing cable, or with the power supplied by a portable generator. An electrically powered tool is lighter and more pleasant to operate than its petrol-driven counterpart but suffers the disadvantage of the input power lead which can be a source of extreme danger as it is easy to cut, chafe or trap during use, thus exposing the live conductors. The agricultural and horticultural industry is one of the few that is still allowed to use portable power tools at standard mains voltage (240 V). Similar tools used in, for example, the construction industry are limited to a 'safe' voltage of 110 V. These tools are supplied from a 240 V supply via an isolating transformer, which is always sited at the supply end of the trailing cable. To avoid using the wrong voltage, different plugs and sockets are used for each voltage.

Portable generators with petrol or diesel engines are now available with both 110 V and 240 V supply sockets. One should regard the 110 V as the only one to be used for portable tools, and leave the 240 V socket for powering fixed mains equipment in the event of supply failure.

Where 240 V power tools are still used, suitable supply circuit protection must be applied. Small tools are 'double insulated': this means that the case is fully insulated, both from the motor and from the blade etc. This latter is important for the protection of an operator should he cut or drill a high-voltage cable.

A double insulated tool has only two wires in the supply cable – 'line' (live) and 'neutral'; there is no 'protective conductor' (earth). Use of a three wire cable with the protective conductor does not guarantee operator safety, and can prove dangerous, as is shown in the following example.

A three core cable is trapped under a sharp object: the protective conductor (earth wire) is severed and the pressure forces the severed end of the protective conductor, which runs to the tool, through the insulation of the line (live) conductor. This will result in the tool case becoming live, but with no current flowing into the socket half of the severed protective conductor to operate the fuse trips.

The normal 'earth leakage trip' used for circuit protection can be rendered ineffectual in the same way and cannot guarantee full protection. The best protection is a 'current operated trip'. This monitors the current in the wires to and from the tool (line and neutral). If everything is correct, then the current in the two wires will be the same. In the event of an earth leak, some of the current that normally returns via the neutral leaks to earth, so causing current imbalance which operates the trip. These devices are available in the form of a 13 A square pin plug adaptor, so that any portable tool can be plugged into the mains through it.

The other danger to power tool users is from water entering connections in the mains lead. Water-resistant and waterproof plugs and sockets are available, but expensive. However the use of one of these, plus a current balance trip, will ensure both safety and freedom from supply lead failure.

1.6.2 Hydraulic power

This has been discussed previously. It is safer than electricity, but the supply pipes are more expensive than cable and not as flexible. Hydraulically driven machines are normally restricted to within a few metres of the power sources. The tractor's oil pressure might be too high for some 'industrial' tools.

1.6.3 Bowden cable

This is a multi-strand steel cable rotating within a flexible tube, the most common example being a car speedometer. The power transmission capabilities are limited and high wear will result from sharp bends in the cable. The system was used for small tools, such as hedge trimmers, as a remote drive from a garden tractor, but has largely been superseded by electric drives.

1.6.4 Pneumatics

Compressed air motors are not commonly used, except in engineering workshops. Certain specialist applications, such as power pruners, utilise compressed air rams.

As a power transmission source, compressed air has many advantages: it is safe, leakage does not contaminate, and only a single small bore tube is required.

2 SOIL PREPARATION AND CULTIVATION OPERATIONS

Many operations are associated with preparation and handling of soil, and other basic media for plant growth. At first sight, cultivation and soil tilling are two of the simpler operations in horticulture and agriculture, but in practice great skill and use of the correct equipment are required. We shall encompass all the likely steps taken to produce a suitable medium for plant growth, and operation of subsequent machinery. At the end of the chapter we will look at the preparation of composts.

The ideal structure (figure 2.1) for horticultural soil is produced by several machine operations. Agricultural cropping, such as for cereals and grass, can often tolerate coarser soil particles and trash on the surface, which enables the operations to be reduced or even combined into one.

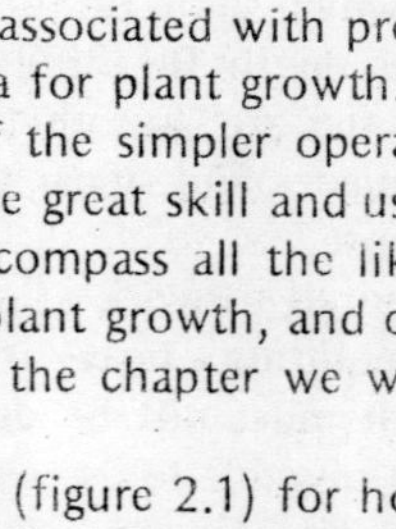

Figure 2.1 Ideal soil structure profile.

2.1 SUBSOIL PREPARATION

This falls into two distinct operations – *moling* and *subsoiling*. While appearing similar, each has its own purpose and machinery.

2.1.1 Moling

A 'mole' is a small tunnel (about 50 mm in diameter) formed in the subsoil to conduct water into the drainage system. Where piped drains with permeable backfill (see chapter 7) are installed, the mole tunnel should run into the backfill. In small areas, the mole can run directly into a drainage ditch. Water from the cultivated layer enters the mole tunnel down the vertical slot caused by the leg. A mole plough and its effect are shown in figure 2.2. To be successful, a mole must fall consistently at a slight gradient (1 in 500 to 1 in 200) towards the drain. Any unevenness accumulates water, so causing it to collapse. To obtain the required depth, the mole cannot normally be used on a tractor lift linkage which varies depth with draught. Instead, a mole plough is affixed to a long frame beam which slides on the soil surface and maintains the mole tunnel at a fixed depth. By using a long beam instead of wheels, minor surface irregularities are smoothed out. The mole bullet and expander are shaped to squeeze and smear the soil, to ensure that the tunnel walls will resist collapse.

A mole plough is often required to work at depths of 500–600 mm – a heavy draught operation requiring a heavy tracklayer or four-wheel drive tractor. For surfaces that must not be destroyed by wheels or tracks, such as sports fields, it is possible to use winch power. In suitable soils, a well-drawn mole can last 10–15 years.

2.1.2 Subsoil loosening

Subsoiling (or under-bursting) differs from moling in that it is designed to shatter the subsoil to allow roots and oxygen, as well as water, to penetrate. This operation is carried out when the subsoil is dry (that is, non-plastic) so that the cracks promoted by the leg and foot travel both laterally and vertically.

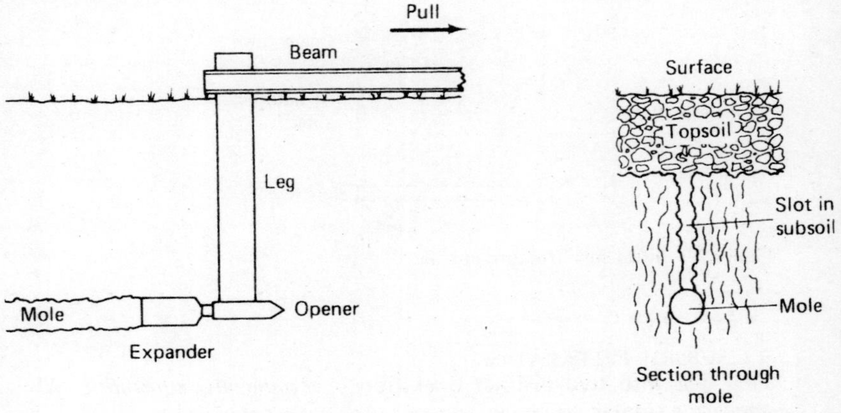

Figure 2.2 Mole plough.

2.1.2.1 Types of subsoiler

The subsoiler is similar to the mole only in that it has a vertical leg with a foot at the base. The foot is different from the mole opener in that it is a 50 mm square section with a chisel point (figure 2.3(a)). The chisel point, acting in dry ground, forces soil upwards as it passes through. The upward heave is translated into the soil above, as shown in figure 2.3(b). The soil cracking is roughly triangular, and thus there are areas between the tine passes where there is no cracking.

Recent developments have been to widen the effective area of a subsoil tine, resulting in a 'winged subsoiler' as shown in figure 2.3(c). The wings as fitted to a normal subsoil tine are 75–100 mm wide on either side of the leg; often ploughshare points are used.

In addition to a normal subsoil tine with wings, models are now available with very wide wings up to 300 mm on either side of the leg. These are designed to ensure that all soil above them is cracked by lifting it. The power requirement of this latter subsurface cultivator is high, and it normally operates at 200–300 mm deep, in contrast to the 300–450 mm of a normal subsoiler. The effect of this machine is a network of vertical cracks, similar to those induced in clay during dry weather. It is ideally suited to remove the surface compaction caused by vehicles, and can be used on sports turf areas that have been compacted by play in wet weather.

Many attempts have been made to apply some of the power for subsoiling operations through the power take-off, rather than all via the wheels as draught. Most have centred on a standard type of subsoiler in which the leg or foot reciprocates or the whole frame is vibrated.

A further method for using independent power for full subsoil cultivation has been developed by Wye College, University of London. It consists of a standard mouldboard plough with a rotary cultivator unit working beneath each plough body.

The slant tine subsoiler has its subsoil leg canted 45°, instead of vertical. This allows it to undercut a block of soil with the normal lateral wedging action of the leg translated partially upward. These legs are mounted on a frame in staggered formation, like plough bodies, so that blocks are cut and lifted sequentially.

2.1.2.2. Subsoiler configuration

As a subsoiler does not have to operate at a precise depth, it can be tractor linkage mounted. A single tine unit can be used on tractors up to 60 kW (80 hp). Larger ones normally have only two tines and, if used with a high-powered tractor, can be pulled at a relatively high speed. The slant tine and very wide winged (shallow subsurface cultivator) units are used in multiples on a frame. These latter types require

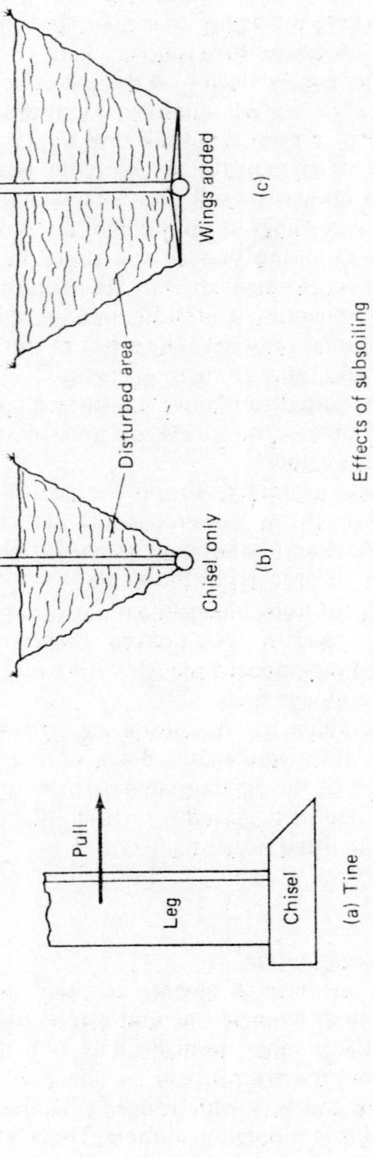

Figure 2.3 Subsoiler.

30–50 kW per tine, depending on soil conditions. The twin leg subsoilers and shallow subsurface cultivator tines are mounted on a heavy toolbar frame with clamps to allow tine spacing to suit conditions. The latter machines often have a vertical pivot in the tine mounting, to accommodate slight sideways deviation without buckling the tine.

The majority of subsoilers have 90° rake (vertical) tines, rather than forward rake which would reduce power consumption. The zero rake prevents large blocks of subsoil being forced to the surface.

2.1.2.3 Subsoiler operation

A subsoiler should operate at a depth that will shatter the greatest amount of soil. This will normally depend on the position of the panned layer, the foot being set to work just below it. If set too deep, it will probably be working in plastic soil, so its wedging action will be nullified. The only reliable method to determine tine depth and spacing is to dig a trench across the line of work. This will indicate the extent of shatter in the panned layer.

Work is being carried out at Silsoe College, Silsoe, Bedfordshire, to determine the optimum subsoiler design that combines the least draught with the maximum amount of soil disturbance. The 'wings' concept, explained earlier, was first evolved here. It was then discovered that the lateral effect of a winged tine depended on the weight of soil above, and that the effective width was reduced if there was too much soil to lift. This has led to an arrangement using two rows of tines, the first row working at approximately half the depth of the second (main) row of tines (figure 2.4).

The effects of using the 'pre-breaking' tine bank have lowered draught or enabled main tine spacings to be widened without loss of effect. The reduced draught has enabled small tractors (40–50 kW) to be used for subsoiling.

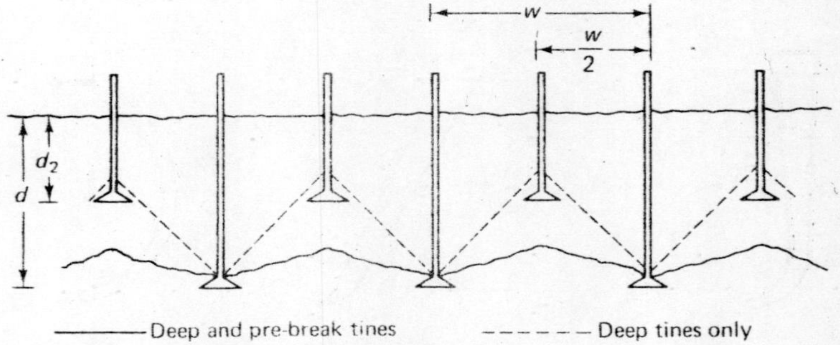

Figure 2.4 Effects of pre-breaker tines.

2.2 PRIMARY CULTIVATION

2.2.1 Surface preparation

Often it is necessary to cultivate the surface of a field to loosen it or destroy unwanted plant material.

This operation can be for one or more of four reasons: surface drainage, surface drying, killing weeds, and encouraging early germination of weed seeds, for which there are three basic machine types.

2.2.1.1 Tined cultivators

Like subsoilers, the cultivator tine has a share point at its lower end which rips into the soil surface. However, the shapes of the tine and its share are designed to tear out the soil above; additionally this geometry allows easy penetration and relatively low draught. The main tine shapes are shown in figure 2.5.

Compacted soil ripped out by the share tends to ride up the tine and be left on the surface. If the soil contains the roots of weeds, such as couch, they will be brought to the surface where the sun can dry them out; also, seed lying on the surface will fall among the clods and germinate.

The tines have a cutting width of 25–100 mm. To ensure that the maximum area of soil is ripped by tines, they are mounted in groups or banks on a frame. The formation of the banks is important, to allow trash and large clods to flow among the tines without blockage (figure 2.6).

Most tined cultivators have some means of avoiding damage due to over-load should the tine strike an obstacle. The simplest is a 'shear bolt', a small bolt in the frame attachment which has been sized to break cleanly and allow the tine to swing freely. As this needs replacing after each breakage, a refinement is a sprung catch which allows the tine to return to its working position after over-load.

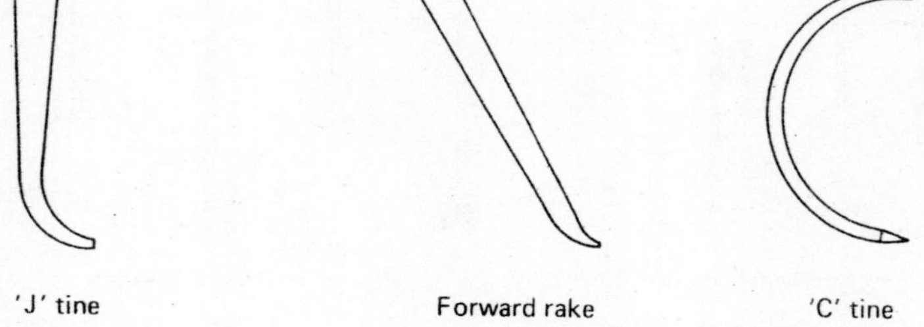

Figure 2.5 Basic tine shapes.

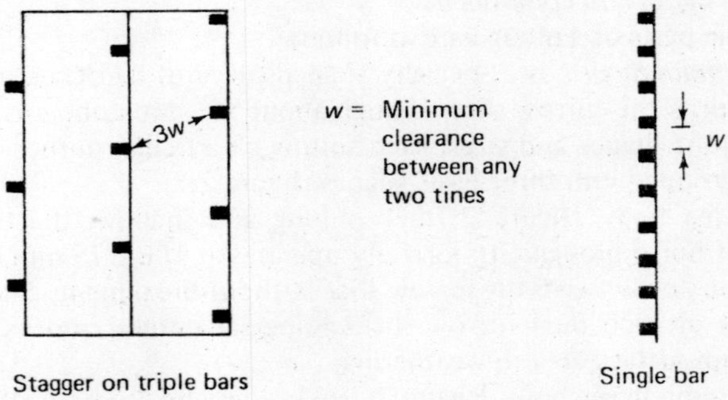

Figure 2.6 Effect of tine stagger.

2.2.1.2 Disc harrows

These consist of a gang of saucer-shaped discs, 250–300 mm diameter. They are drawn across the surface at an angle to cut out a slice of soil. The angle can be varied, wide for maximum penetration, narrow for maximum lateral disturbance. The disc gangs are always assembled from an even number of disc shafts, with each at opposing angles to balance side thrust from the disc.

To aid slicing action on surface trash, discs on the first shaft often have notched edges. Further designs use a 'disc' consisting of four arms which tend to pluck rather than slice the soil surface.

2.2.1.3 Rotary cultivators

A horizontal rotor machine is described later. This utilises the power source directly, rather than hauling it along the surface.

A variation on this type uses two ground-driven rotors with thin blades. The rotors are geared together so that one revolves faster than the other, in a combination of tine and horizontal rotor action.

2.2.1.4 Vegetation destruction

The tine cultivator can be used to break out root vegetation but has little effect on above-ground crop residues. The disc harrow or rotary cultivator is ideal for mulching standing crop residues, the rotary one dealing adequately with large hard brussels sprout stems.

2.2.2 Ploughing

This traditional tool is still the mainstay of operations, although on small nursery areas it has been superseded by the rotary cultivator.

2.2.2.1 *Types and components*

The basic parts of a plough are as follows.

The *mouldboard* is a specially shaped piece of hardened alloy steel which turns the furrow slice through about 90°. Mouldboards are made in different shapes and sizes, each suiting a particular purpose but they can be grouped into three basic shapes (figure 2.7).

The lea body (figure 2.7(a)) is long and shallow, the traditional shape of horse ploughs. It normally operates at 100–175 mm deep, and the shape slowly twists the furrow slice without breaking it. The finished effect is of individual furrow slices lying separately, and exposes the maximum surface area to weathering.

The semi-digger body (figure 2.7(b)) is designed to work at a greater depth than the lea, up to 250 mm. It is deeper in section and shorter; this does not keep the furrow slice intact, and the finished effect is more broken with less distinct furrow slices.

The digger body (figure 2.7(c)) is designed to lift soil from 300 mm or more. Its shaping is very abrupt compared with the lea, and it produces a well-pulverised slice.

Modern ploughing practice dictates a flat finished, well-broken furrow slice, hence most ploughs are fitted with mouldboards shaped somewhere between semi-digger and digger. These have been designed to plough at higher speeds, 5–8 km/h, commensurate with modern tractor powers.

For soils that do not slide freely on the mouldboard surface, it is possible to fit slotted or skeletal mouldboards to lessen the surface area for cohesion.

The *share* is mounted at the front end of the mouldboard, and cuts the furrow slice. It is subject to hard wear and is thus a replaceable item, separate from the mouldboard. The share design varies with the mouldboard.

The multi-piece share allows for the parts to be replaced according to the severity of wear.

Some ploughs for very hard or stony land have a share point in the form of a spring-loaded steel bar projecting through the mouldboard's nose (figure 2.8). As the 'bar point' wears it can be adjusted forward, and if it strikes an obstacle it will retract and allow the plough to ride over it.

Coulters perform a partial vertical cut, so that the furrow slice is broken out neatly by the mouldboard.

The knife coulter is a vertical blade, cheaper than disc coulters, but it will not cut surface trash. Its slight rake allows stones to ride out without damage.

The disc coulter is a thin steel disc about 250 mm in diameter, mounted on a castor action arm to accommodate slight sideways movements in the plough without it buckling. The disc is much more able to

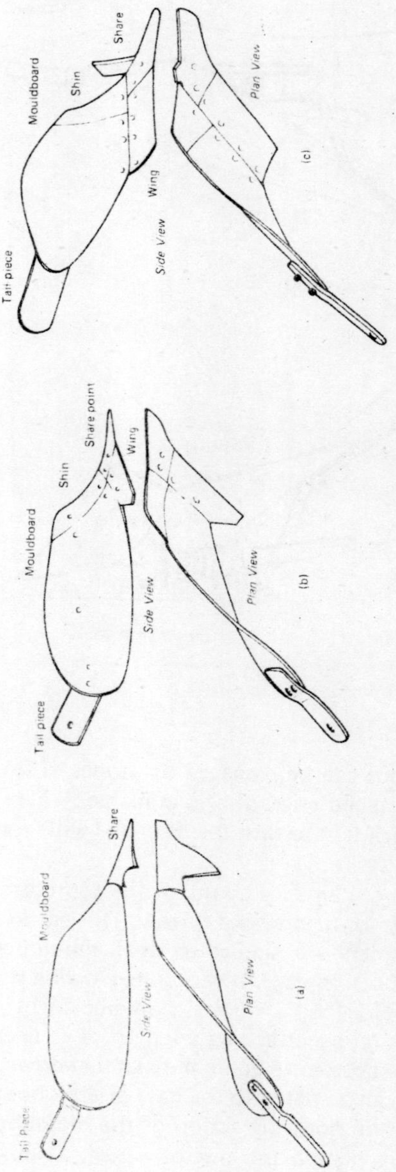

Figure 2.7 Plough body shapes: (a) lea body, (b) semi-digger body, (c) digger body.

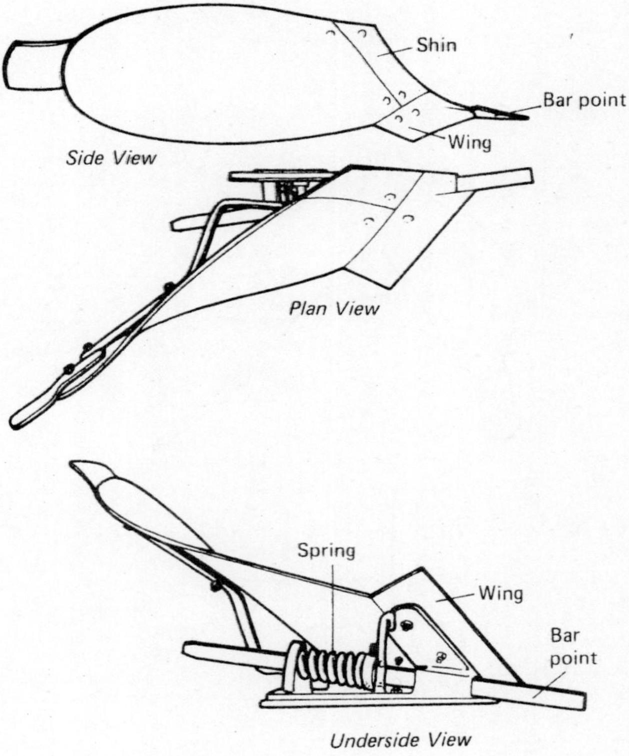

Figure 2.8 Bar point body.

cut surface trash, but can be damaged by stones. If surface trash is too tough to cut, a notched edge disc is available. Where the soil surface offers insufficient friction to turn the disc, one with a wavy edge can be fitted.

Skimmers remove a small amount of the top corner of the furrow slice, and throw it into the previous furrow. This ensures that all trash is buried. Most skimmers are similar to small mouldboards, and might have a replaceable bottom edge. If very long growing vegetation, such as mustard, is being ploughed a normal skimmer might not bury it. The answer is often found by attaching a length of chain from the plough beam to the furrow bottom to guide the material down.

Subsoiler tines can be fitted to the base of each body to loosen compaction caused by the smearing action of the mouldboard base. In the furrow immediately behind the tractor wheel, a subsoil tine removes weight compaction. Many modern ploughs with light frames do not have the strength for a full set of subsoilers. If only one is fitted, it should be in the wheeling furrow.

Researchers at London University's Wye College have developed a 'double digger' plough which incorporates a rotary cultivator head beneath the mouldboard; this is designed to plough and fully subsoil in one operation.

The plough has two basic configurations, single way or reversible.

The single way plough has one or more mouldboards, all turning the soil to the right; thus it can operate only in a set pattern although, even so, double furrows are occasionally left where blocks of ploughing meet.

To overcome having to work in blocks, the reversible plough was designed. This has a set of right-hand and left-hand bodies with the facility to put either into operation, which allows ploughing to start at one side of a field and progress across with the plough shuttling to and fro against the previous furrow.

Small single furrow, single way ploughs can be fitted to walking garden tractors. Conventional tractors can take ploughs of one or more furrows, single way or reversible. These are normally three point linkage mounted up to four or five furrows, and semi-mounted if larger. Semi-mounting involves mounting the front end on the tractor's lower lift arms, with the rear end supported by trailed wheels.

The latest innovation is to mount a small pushed plough on the front of a four-wheel drive tractor, and a conventional one behind. This arrangement is considered to provide better weight for traction on the front axle and to improve performance.

Many plough frames are now made so that the rear body can be removed to allow operation in poor conditions.

Disc ploughs have the mouldboard share and coulter replaced with one large disc. This disc is of 750–900 mm diameter and slightly dished. It is angled so that as it rolls through the soil it cuts and turns the furrow slice. The turning action is much more random and does not fully bury trash. Its main uses are on thin soils where it can roll across stone, and where surface trash is desirable to protect against soil erosion.

2.2.2.2 Plough setting

Correct setting of a plough should not be regarded as solely to obtain an aesthetically perfect finish; a correctly set plough requires less draught, and properly performs functions such as trash burial. The basic points in plough setting are

(a) The tractor wheel settings must be correct for the width of plough used. For reversible ploughs the distance from the tractor's centre line to the wheels must be the same for both sides.

(b) The plough frame must be level when in work. Longitudinal level (pitch) is achieved by adjusting the top link, or depth wheel if semi-mounted. On single direction ploughs, transverse levelling is by the adjustable three point linkage lift rod. On reversible ploughs the linkage lift rods must be of equal length, and transverse levelling is done by adjusting the appropriate roll over holding latch for each side.

(c) Very few modern ploughs have adjusters on individual bodies. Where these are fitted, setting up is carried out on a flat floor, following the manufacturer's instructions; they are not adjusted during work.

(d) Coulters should cut about 12–20 mm outside the line that the mouldboard side will sever, to ensure a clean furrow wall.

(e) Skimmers should cut off the corner of the furrow slice as it begins to rise up the mouldboard. They should not be set to turn out a small furrow ahead of the mouldboard.

(f) The plough will find its own running position laterally behind the tractor. This will come from the balance between the forces acting on the plough, and the direction of pull from the tractor. It is wrong to restrain lateral movement with the link arm stabilisers, or manual effort if on a garden tractor. The lateral position relative to the tractor determines the width of the first furrow; this is adjusted by altering the angle of the plough relative to its mounting on the tractor linkage. On the single way plough this is achieved by rotating a cranked cross-shaft, which carries the lower linkage attachment pins. The reversible plough is adjusted by altering the angle between the plough frame and the mounting headstock, in order to effect the same degree of adjustment to both sides.

Setting a single furrow plough is often harder than setting a multi-furrow plough. A well-set plough will virtually steer itself while a badly set one will prove impossible to keep straight.

2.2.3 Rotary cultivators

These utilise mechanical power directly for tilling rather than as a draught force pulling a body through the soil.

The machines in this category are based on a horizontal rotor fitted with blades. On most machines the blades are 'L' shaped, although other configurations, including hooked and straight tines, are provided by some manufacturers. The blades are staggered around the rotor to even the torque loading.

The rotor rotates in the direction of travel, soil resistance on the blades causing the machine to drive itself forward, and this has to be restrained. On large pedestrian-controlled and tractor-mounted machines. restraint is through the driving wheels. Small pedestrian machines are restrained by a fixed tine drawn through the soil below digging depth, its depth determining the digging depth. Set deep, the rotor needs the resistance of a large depth of soil to provide forward movement; set shallow, the rotor can find sufficient forward thrust in a small depth of soil. In use, bearing down on the handles forces in the restraining tine so the machine moves forward slowly, digging deeply; raising them allows a fast forward speed with less depth.

Normally, wheel-driven pedestrian and tractor-mounted machines do not have restraining tines. Some large tractor-drawn machines have a bank of fixed tines at the rear. These are used to break the soil below rotor-digging depth, utilising the forward thrust from the rotor to provide a large part of their draught.

The combination of rotor and travel speed determines the size of each slice that the blade cuts. For primary cultivation it is normal to use a low rotor speed combined with a high travel speed to break the ground into large lumps which will not slump under winter rains and negate weathering effects. Research at the N.I.A.E. has developed a 'rotary digger' on the horizontal rotor principle, which has a slow speed rotor, fitted with long blades, to break the soil into large lumps. The 'L' blade is shaped so that there is a clearance between its underside and the soil, to avoid smearing at most combinations of rotor and forward speed. It is possible however for the cutting edge to polish very wet or very dry soil, if a high rotor speed is used with a slow forward speed.

Rotary cultivators do not fully bury trash, although long material will be chopped. They are used to chop crop residues, such as brussels sprout stems, often as a pre-treatment for ploughing.

Certain designs of rotary cultivator have blade assemblies which are mechanically twisted as the rotor turns. The twisting action resembles the movements of a spade blade while digging, and they are often termed 'spading machines'. They are mechanically complex and slow in operation but have proved useful in glasshouses where inversion of the soil and trash burying are important.

2.3 SECONDARY CULTIVATION

This covers operations that convert primary cultivated land into a smooth, level seedbed for sowing or transplanting. The objective is a fine crumb structure, level soil surface and even compaction; the machinery should not destroy soil structure produced naturally by weatheing. It is also important to prevent the wheels of cultivation machinery causing areas of high compaction where the crop is to grow. Tilth requirements differ with the cropping, ranging from a loose tilth under root vegetables, to an even, well-compacted seedbed for establishing amenity grass.

2.3.1 Tilth

A roughly cultivated surface left over winter will consist of a layer of fine crumbly soil overlying wet lumpy material. During cultivation the weathered crumb must be preserved on top without recompacting or burying it among the unweathered lumps. The basis of secondary cultivation equipment must be to loosen the soil in its respective layers without pulling unweathered soil to the surface. This is mostly done with tines, the shape of which is important. A simple rule is to avoid tines with curved bottom ends or set at an angle, both of which will cause soil lumps to climb up as they pass through. Likewise, a rotary cultivator blade will vertically mix soil, and can cause unweathered clods to be left on the surface. This has been overcome somewhat with the use of tines rather than blades which do not cause so much soil lifting.

On a well-weathered soil the clods will crumble easily from the stress caused by tines jostling them, or the rolling pressure (figure 2.9).

Most operations involving simple tines require two or more passes to achieve a suitable tilth. The second pass is often at right angles to the first, which improves the cultivating action but results in a greater area of the field being subjected to tractor wheelings.

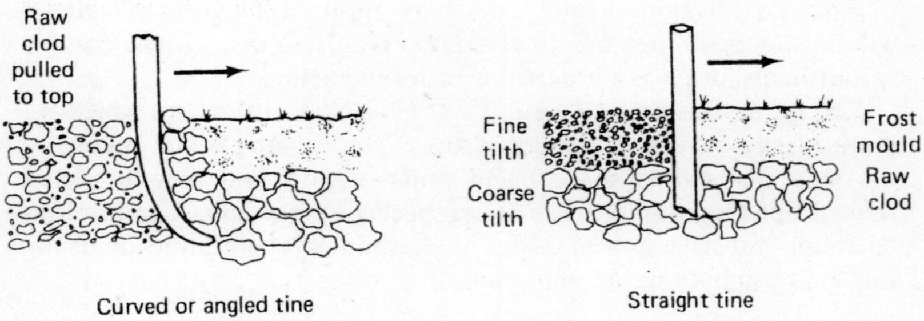

Figure 2.9 Effect of tine shape on tilth.

2.3.1.1 Rigid tine

The 'dutch harrow' consists of a heavy timber or box steel section frame with short straight tines projecting beneath. In operation the frame runs on the ground and aids the breakdown caused by the tines by means of its scrubbing action. The frame bars also do some surface levelling. The tines are clamped into the frame by bolts, so that their projection can be adjusted. Power consumption is relatively low, and a 45 kW tractor should be able to handle a 4 m wide dutch harrow.

2.3.1.2 Reciprocating tine

The reciprocating harrow, consisting of tined bars which reciprocate at 90° to the direction of travel, has the same action as a simple dutch harrow, except that part of its power requirement is supplied by means of the PTO, rather than being all draught. The reciprocating action enables the machine to produce the same effect in one pass that requires two or three passes with the non-powered version. There is always an even number of reciprocating bars, paired on the drive crank or wobble box, so that one of each pair moves in the opposite direction to the other to balance side forces. Normally machines have only one pair of bars, although some are equipped with two pairs (four bars).

2.3.1.3 Rotating tine

Vertical axis rotary tined cultivators are increasingly being used to create tilth in one pass under most conditions. The tines are mounted in twos or fours on vertically rotating units which are geared to rotate together. Most of the power consumed is through the PTO, and the total requirement is higher than for the dutch or reciprocating harrow, needing 75 kW for a 4 m width.

2.3.1.4 Spring tine

Spring tined cultivators, with flexible tines having a curved lower end, produce a loosened seedbed but bring clod to the surface. Spring tines are mostly used for cereal seedbeds which can tolerate rough sowing conditions, and where clods are unimportant to successive machines. They might have a use where roughly cultivated land needs to be pulled around in dry weather to kill weed infestations.

2.3.1.5 Rotary cultivators

Certain of the horizontal rotor machines used for primary cultivations can be used for seedbed preparation also, the only change being to combine a high rotor speed with a low travel speed. On pedestrian-controlled machines this involves a deeper restraint tine setting to hold back its forward progress. On machines with a gearbox or multi-sized drive pulleys, the highest ratio is chosen. Tractor-mounted machines are used in a low forward gear; most models also have a selection of PTO drive gear ratios to allow a range of rotor speeds at standard PTO speed. For secondary cultivation alone, lighter, wider rotary cultivavors with thinner blades which reduce vertical mixing are available. A better method to reduce vertical mixing and smearing of lower layers is to fit the rotor with tines instead of blades. These are normally used at the highest available rotor speed.

2.3.1.6 Machines for consolidating tilth

Tilth left by rotary (horizontal and vertical) machines may be too loose for good drilling or planting, so many of these cultivators are fitted with a trailing skeletal roller which slightly recompacts the loose tilth. On many machines this 'crumbler roller' also acts as the support wheel for depth adjustment.

Where soils are easily broken down and liable to produce a very loose tilth, on peatland for example, a cultivator consisting of roller shafts fitted with small discs (approximately 150 mm diameter) at 100 mm intervals is used. This is commonly termed 'fen press'. The combined action of shaft and disc breaks and levels the surface without lifting or loosening it.

2.3.1.7 Rolling-action tilth cultivators

The most common of these is the disc harrow, as previously described. This type of cultivator is not normally considered suitable to provide a tilth for root crops, as the discs tend to compact and smear the soil beneath them. They can also tend to peel up strips of unweathered soil, and mix them into the tilth. They could be suitable for establishing grass in amenity horticulture where the compacting action provides a top tilth suitable for seeding and rolling, and a firm subsoil which resists sinking when the grass is established. Their ability to cut into firm unweathered soil can also make them useful to alternate with a land levelling plane in preparing level pitches.

The use of notched rollers and wheels is preferable in some situations. Although in certain soil conditions the stress to break weakened clods can be provided by weight, a plain or ribbed roller will not provide sufficient stress or penetration of that stress, so rollers and wheels with nobbles or projections have been designed to produce localised areas of high stress to break clods. The danger of this type of machine is in wet conditions, where the nobbles will cause the surface to puddle and compact.

2.3.2 Tilth damage

2.3.2.1 Tilth problems

In many instances problems in crop establishment can be traced to soil structure damage caused during cultivation. This can manifest itself as

smearing or panning – an impervious layer formed within rooting depth;
slumping – loss of previous structure when soil is wetted

capping	– impervious surface layer caused by heavy water drops from rain or irrigation
clod	– compacted or unweathered pieces of soil in tilth layer
'fluffy' tilth	– soil is too loose to provide good conditions for plant development.

2.3.2.2 Avoiding damage

Damage can often be avoided by correct choice and use of machine. The following points will be helpful

(a) Cultivate only when the soil is dry enough to crumble; if too wet, it will pack and form clods. Be careful of soil with a dry surface which might be too wet lower down.
(b) The tine ends of vertical rotor cultivators and the 'L' blades of horizontal rotor machines can smear wet soil at the bottom of their working depth.
(c) Rolling action of discs or wheels will compact (pan).
(d) Slumping and capping are often caused in a tilth that is too fine; leaving small clods might prevent this. Soils that are known to slump should be cultivated immediately prior to sowing or planting.
(e) Avoid machines that mix vertically and bring up unweathered soil.
(f) There are times when soil will break easily, others when it will not do so without great effort. The present generation of high-powered tractors and cultivators should be used under the right conditions, rather than to beat a soil into submission under adverse conditions.

2.3.3 Ridging

Many horticultural crops are grown in ridges or require soil to be earthed around while growing. In Scotland most root crops are grown on pre-formed ridges. Ridgers can be divided into four types.

2.3.3.1 Mouldboard

This is similar to the mouldboard plough, in which a pair of opposing boards move the soil sideways to form the ridge. There are two basic shapes of mouldboard; one similar to the lea body, the other to the digger body.

In the lea type the soil is lifted from the furrow by a share point and slides along the body to form the ridge side. The share can penetrate firm soil in the furrow but this can lead to clods in the ridge; the body shape can also smear the ridge sides if the soil is damp.

The digger type is shorter than the lea and set to a more abrupt angle. This results in loose tilth being deflected sideways, without smearing. The abrupt operating angle is likely to lead to a build-up of soil on the board face in wet, sticky soil, which can cause high draught.

The lateral forces created by the mouldboard can cause narrowing of crops planted in wide bands, such as narcissus bulbs.

2.3.3.2 Disc

The disc ridger uses two large, dished discs which rotate as they pass through the ground, and lift and push the soil sideways. Discs will make only a relatively low ridge, and are often used as a primary ridging tool, followed by a mouldboard or powered machine. Discs have less draught and cause little smear on the ridge sides, but can slice raw soil to form a cloddy ridge.

2.3.3.3 Rotary

Two types of powered ridger can be found. One is a combination of mouldboards following a narrow rotary cultivator. The cultivator loosens tilth in the furrow bottom, and the mouldboard guides the soil thrown up by the blades into ridges. This can reduce the smearing action of the fixed mouldboard type. Alternatively the rotary ridger uses inter-row horizontal rotors with blades or tines in a helical pattern, which cultivate and move the soil sideways in one operation. The resulting ridge is formed of very loose friable soil, and is the least likely to contain clods; this is an advantage for mechanical harvesting. Both types contain complex power drives which complicate facilities for altering row width, and some machines are not adjustable.

2.3.3.4 Multi-pass

The foregoing techniques form the ridge in one or two passes. A system exists for forming clod-free ridges and cultivating for weed control. This involves several passes with banks of small tines running between each row. The tine banks are arranged so that one set cultivates the furrow bottom, another set cultivates the ridge sides to remove weeds, and the final set is angled to reform the ridge. Their action should, in theory, only form a ridge with tilth that has weathered since the last pass. This system is being superseded by one pass systems to allow the use of surface-acting herbicides for weed control.

2.3.4 Bed systems

Horticulturists are concerned to minimise compaction in the area of soil in which the plant grows. Most soil compaction occurs from tractor wheels when cultivating to form the seedbed. Although low-pressure tyres or tracks can help, it is possible to eliminate this damage only by segregating tractor wheeling paths from crop areas at all stages of cultivation and growing. This results in a series of parallel wheeling paths with growing 'beds' between. These beds are established with the first machine to pass after the plough, often an empty tractor fitted with marker guides. The normal wheel-centre for bed work is 1.8 m which gives around 1.5 m of unwheeled soil between. Cultivation is as previously described, except that it can be carried out in one direction only, instead of by multiple passes at different angles. The cultivator must also work only the bed soil, to avoid digging compacted soil from the wheelings, and pushing the resulting clods into the bed. With small operations it is normal to cultivate only the bed immediately behind the tractor with a 1.5 m wide tool. Large-scale growers demand greater working widths, so it is common to see three 1.5 m tools on a frame behind the tractor, numbers 1 and 3 cultivating beds either side of the tractor, and the machines being spaced to avoid the wheelings.

A further development of the bed is as a replacement for two or three individual root crop ridges. In theory the better spatial arrangement will result in higher yields, but this will require different techniques in planting, bed covering and harvesting.

Research is being carried out into permanent beds where the wheelways are never cultivated, and are even stoned or concreted. This involves a full set of bed-sized tackle from primary cultivation to harvest. Initial observations indicate that many operations are easier owing to uncompacted soil, and require lighter machinery.

A further development where the wheelways are 12 m or more apart requires all machines to be mounted on a large travelling gantry frame which can span the bed. Most machinery will be mounted on the gantry so that it can move laterally (along the gantry) to sequentially work the bed in strips, rather than having to cope with the whole width.

2.3.5 Stone treatments

This technique is now widely practised in stony soils growing root crops. It is also used in amenity horticulture where stony sites are to be grassed or tidied.

2.3.5.1 Windrowing

For root crops, the normal method is to move the stone away from the immediate crop area only, normally into the wheeling tracks, prior to planting. The machines resemble potato harvesters, but are more strongly built. A complete band of soil about 300 mm deep is dug; soil and small stones fall through the web bars on to the ground, large stones (normally over 40 mm) and clods being retained on the web. It is impractical to remove these stones from the field because of the quantity involved (this can be as much as 750 t/ha within cultivation depth), so they are deposited where they can cause no problems.

Two systems can be used, depending on whether the crop is to be planted simultaneously with destoning, or as a subsequent operation.

In the first method the planter directly follows the destoner, or is incorporated within it. The collected stones are deposited into a furrow between the ridges planted at the previous pass. There is often no need for further ridging operations and thus little risk of re-introducing stones from these furrows. The discharge conveyor is reversible to allow two way working.

In the second system a deep furrow is formed behind the destoner wheel to receive the stones. These have to be placed sufficiently deep for subsequent planting and ridging machines to clear them. The placement ploughs add to the power requirement of this system.

Some destoners have a secondary separator to retain stones over 125 mm in a hopper for carting off the field. Work rates can be up to 4 ha/day, and power requirements 50 kW–75 kW.

2.3.5.2 Crushing

A theoretically more permanent remedy is to crush the stone to a size that will not present problems. Machines have been built to do this using rollers or impact mills. Results have varied – much depends on the type and cleanliness of the stone – and the machines absorb high power and give low output. Some stone types break into sharp-edged fragments which cause heavy surface damage to crops as they are harvested.

2.3.5.3 Removal

Amenity site preparation, especially where it involves coal tips or quarries, can involve removal of large surface stones prior to seeding. This requires a different approach in that the only stone of consequence lies on the surface; although this is a small quantity, it has to be removed completely.

The simplest system uses a rake running at a 30° angle to travel, with fixed tines or a contra-rotating tined rotor. It can work down to 100 mm, leaving stones in windrows, from which they can be cleared by a stone picker machine or excavator bucket. Another type of rake-operated stone remover has a bank of rake tines running at an angle to the soil surface. Large stones caught on the tines are periodically transferred into a hopper by tipping the whole rake bank backwards into it.

Stone picker machines use powered rakes to deflect stone on to a pick-up conveyor, to transfer them directly to a hopper or trailer. One model uses a large rotary drum elevator to avoid the wearing of parts that is encountered with chain conveyors. The output of these machines is 1–2 ha/day, dependent on the stone population.

2.3.6 Light harrows

These are used for final levelling of tilth or raking soil over broadcast seeds. Two types exist, rigid frame and chain. The former is a lattice of bars carrying shallow tines about 100 mm long. The 'chain' type is a matrix of steel rods, bent to hook together with their free ends pointing downwards. The chain harrow is better able to follow undulating ground, and is used also to rake established grass.

2.4 HOEING AND INTER-ROW CULTIVATIONS

Three types of hoe are used to destroy weeds between rows of growing crops.

2.4.1 Fixed blade

The fixed blade hoe consists of thin steel blades which run just beneath the soil surface. The blades are commonly 'L' shaped, a right-hand and a left-hand one being on either side of each row, with the vertical shank next to the crop. If the inter-row space is wider than the 'L' blades can cover, an 'A'-shaped blade is set to run between them. To be effective, the hoe should run so as to sever weed roots at about 10–15 mm beneath the surface. If set too deep, the soil will flow over the blades, leaving the weed roots intact. Hoe blades can be mounted on a simple hand-propelled frame or a pedestrian garden tractor for small areas. Large areas use tractor-pulled hoes which used to be fitted to the rear and required a separate operator to steer the hoe between the rows. Modern hoes are mounted ahead or amidships of the tractor to allow the tractor driver sufficient vision for accurate steering.

For some operations side shields are fitted to hold back crop leaves, and prevent damage by hoe blades. Flat discs are available for fitting to run beside the crop row, to sever runners in crops like strawberries. Angled concave discs are used to take soil away from alongside the crop row so that it stands proud; this eases chopping out with a hand hoe for single thin line drilled crops.

2.4.2 Rotary blade

An alternative to the fixed blade hoe is one with powered rotating blades. This resembles small rotary cultivator units which fit between the crop rows. In addition to weed destruction, these units allow limited recultivation and incorporation of chemicals. The mechanical drive adds complexity to the hoe, especially as the units have to be adjustable to different row spacings and cultivating widths.

2.4.3 Ground-driven rolling cultivator

This consists of a number of small wheel units with curved tines. The wheels roll across the soil at a slight angle to the direction of travel, and the tines penetrate the ground surface and lift out the weeds. These machines can be angled to move soil towards or away from the crop row, and are most effective in dry soils when running at a high speed.

2.5 ROLLS

These are used to consolidate the soil, crush clods and help to prepare a smooth working surface. For cereal and grass crops a roll consisting of individual ribbed ribs, commonly called a Cambridge roll, is used. The concentration of weight under the ribs makes it better for shattering large clods, and the multiple small furrows it leaves can be useful for the shallow incorporation of broadcast grass seed.

A smooth surface can be obtained only with a flat roll, available in a selection of weights and diameters. Some are formed from a watertight cylinder which allows their weight to be varied by the addition of water. A large diameter roll is essential for soft soil compaction, as the sinkage on a small roller will cause bulldozing and make it hard to pull.

A very heavy roll is an essential spring treatment for grass growing on stony soils. Here harrowing or spiking might pull stones to the surface which will damage mowers. Rolling will push them back into the surface.

2.6 PREPARATION OF SOILS AND COMPOSTS FOR HORTICULTURAL USE

The main operations in this field are screening and mixing.

2.6.1 Screening

Most of today's horticultural composts are based on organic fibres, such as peat or bark, mixed with inorganic filler, such as sand, vermiculite or polystyrene. If these ingredients are bought in their raw state, the first operation is screening to take out large lumps and long fibres. Screens can range from simple circular hand sieves to rotary barrels or vibrating screens. The screen size chosen should reflect the job that the compost is to do. Fine screens will be needed for composts to be used to fill cellular propagating trays, whereas very coarse material is suitable for 3 litre shrub containers.

For coarse lumpy peat or bark, a rotary barrel screen is preferable. The tumbling action will help to break down lumps, and lessen overgrade waste.

2.6.2 Mixing

The need for thorough mixing is important because of the number of pesticides and nutrients that require to be incorporated. With a poorly mixed compost some pots or blocks will contain too little of the additives and some too much, with resulting toxic effects.

2.6.2.1 Turning

Tipping ingredients in layers on to a heap, and turning it by hand, is still a practical method. In fact, the standard with which animal feed mixers are compared is a heap moved three times by hand shovelling. Given sufficient concrete surfacing, a heap can be turned by tractor loader shovel. This is best done by driving into the heap to pick up a shovelful, then continuing forward to deposit it across the far side of the heap. Turning with a rotary cultivator is not regarded as satisfactory.

Turning is also an essential process during the production of mushroom compost, to maintain the aerobic state of the heap, so that the correct microbial action can occur. In modern mushroom production the compost is held in the form of a long heap, 1.5–2.5 m wide and 2–3 m high. The turning machine is a self-propelled mobile unit, which straddles the heap. Turning is effected by a system of tined conveyors or drums, which break down the heap ahead of the turner and reform it behind. The turner can be either of the 'through flow' type, in which the compost is moved through the heap by the momentum imparted by the tined drums, or the 'conveyor type', in which the material is carried through. In either type the heap is reformed inside large guide plates, which in some cases can be adjusted to alter the heap size. The turner can also form the initial heap from raw compost tipped ahead of it.

2.6.2.2 Batch

There are many machines for mechanical batch mixing, a popular one being the rotating drum concrete mixer. The resultant mix is often uneven, as there is insufficient stirring action to break up masses of peat and other light materials.

There are two types of purpose-built mixer. One is a sloping rectangular box, holding between 0.5 and 1.5 m^3 of compost. Mixing action is provided by an open chain and slat conveyor which runs inside the box with a cross-auger to provide sideways agitation. One end of the box rests on the ground and has the filling hatch; it discharges at the opposite end, which is raised on legs to go over a barrow or conveyor. The filling hatch size normally precludes putting in whole, unbroken peat bales.

The other type consists of a horizontal trough with either one or two horizontal paddle shafts running end to end. The paddles have a helical form so that material is moved laterally while it is being stirred. The mixer is emptied by either mounting it high enough for gravity discharge through a trap in the base, or fitting an inclined emptying auger with sufficient height to fill a blocker or potting machine. This type of mixer is very heavily built, in order to withstand the forces exerted during the direct mixing of whole bales of dry peat.

Most mixers incorporate water injection systems, so that the compost can be finished to the desired moisture content. Some watering systems incorporate a meter so that pre-determined quantities can be added to ensure consistency between batches.

2.6.2.3 Auger

One type of compost 'mixer' consists of an inclined auger with water injection facilities. As there is no means for compost to be recycled, it must be regarded only as a wetting machine for pre-mixed composts.

2.7 WEAR OF SOIL-ENGAGING COMPONENTS

This is of great concern to anyone working with stoney or abrasive soils. Replacement cost is not the only factor; often the time lost while replacing worn parts is of greater significance. Wear can be reduced by four methods.

2.7.1 Component materials

Specially cast steels are commonly used, but unfortunately an increase in hardness often results in a more brittle material, which will snap during use. Therefore wearing-part material is often a compromise between hardness and ductility (the ability to bend without snapping). Some parts are produced as 'chill castings' where the wearing surface is cooled rapidly in the mould to produce a hard grain structure, while the core cools slowly to produce a softer, but more ductile, centre.

2.7.2 Hard facing materials

These are special alloys deposited on to the normal steel surface by welding, the most common being stellite. It is supplied in electrode rods ready for use in an electric arc welder. The material is applied in runs 10–15 mm wide, and can be used on edges as well as flat surfaces. To avoid the need for large deposits on wide surfaces, a grid of weld lines can be drawn at 20–30 mm intervals. Soil sticks in the pockets between the weld lines, and the main soil flow slides over it without touching the parent metal; this does however increase draught forces.

2.7.3 Tungsten carbide

This is a mixture of powders containing tungsten and carbon which are pressed into shape and kiln-baked into a solid mass. The resulting product is very wear-resistant but brittle, and has to be supported by being welded to steel. The expense of tungsten metal combined with the high-powered presses needed to form it mean that tungsten carbide tips are expensive, and are often confined to small items of simple shape.

2.7.4 Ceramics

Mineral oxides can be formed and kiln-fired in a similar way to tungsten carbide. The resulting material is much harder than tungsten carbide but has virtually no ductile strength. Much work has been done at the N.I.A.E. on suitable materials, shape and mounting methods. The latter is extremely important, as ceramics have to be fully mounted on to steel for support. As even the normal flexing of a steel tine or share under stress can break the ceramics, very heavy section steel backing is needed. This results in a fatter share profile, which can increase draught. The ceramic can be glue-bonded or have mounting grooves which match tongues on the steel backing. Although this latter method allows broken or worn ceramics to be replaced easily, it demands a high standard of manufacture of both items. At present, ceramics can last around four times as long as hardened steel, and research is progressing to find a viable commercial system.

3 CROP ESTABLISHMENT

All plant material has to be established, whether from seeds, corms, tubers or as immature plants. Any mechanisms for establishment must fulfil the following criteria

The crop must be established at the correct population.
There must be no damage to the plant material.
The crop must be established at the correct depth.
Care should be taken to avoid destroying soil structure.

3.1 SOWING DRY SEEDS

Equipment in this category can be divided into three basic types

Broadcasting – to distribute seeds over an area.
Thin line sowing – to randomly distribute seeds within a row or narrow band.
Spacing or Precision – to place individual seeds at pre-determined positions.

Some examples of the suitability of sowing methods for different crops are shown in table 3.1.

Table 3.1

	Sowing method		
	Broadcast	*Thin line*	*Precision*
Normal method for	amenity grass	salad onions, radish, sweetcorn, transplants, baby beets, bunching carrots	brassicas, onions, most root vegetables
Can be used for	salad onions, radish, transplants (small areas only)	amenity grass (rows will tiller sideways in time)	any other crop, but speed is slow

3.1.1 Broadcasting

Many of the machines for broadcasting fertilisers (chapter 4) can be used for seeds. The main difference is the reduction of metering rate, as seed is applied at a lower rate than fertiliser. Some machines that rely on throwing material (spinning disc or pendulum) might have a reduced bout width for light seeds such as grass.

One small centrifugal spreader has been designed especially for seeds and pesticide pellets at low rates. The hopper holds 5-10 kg of seed, and discharges directly on to a small disc drive by a 12 V d.c. motor. The spinner unit is small enough to mount high on a vehicle to increase spread width, and it can be driven from the battery without the constraints of power shafts.

Several full width hopper broadcasters have been built for seed. The simplest consists of a V-shaped hopper, 2.5–4 m long, with a discharge mechanism. Often this is a series of small holes to allow the seed to trickle out. Flow is encouraged by a shaft carrying agitators or by brushes running in the hopper bottom. The sowing rate is governed by the agitator shaft speed, and the size of the outlet hole.

On some broadcasters, the seed streams fall on to angled plates to spread them uniformly. Most full width broadcasters are now tractor-mounted; hand-pushed types were once common but are now becoming obsolete.

More complex broadcasters use the same metering mechanisms as the thin line drills described in the next section.

3.1.2 Thin line

3.1.2.1 Cereal drill based

As this is the technique used for cereal seeds, many cereal seed drills are suitable for sowing grass and vegetables. The main limitation to using cereal drills is seed rate, which for cereals can be 10 times that for vegetables. Where the drill cannot be calibrated to a suitably low rate, it is possible to dilute seed by mixing it with inert material, such as limestone granules or vermiculite, to give a volume that can be metered.

The metering mechanisms are referred to as 'force feed', meaning that the seed does not rely on flowing by gravity. The seed is metered by either a hard roller with a pocketed or fluted surface, or by a soft roller which traps the seed against a flat plate. Rate is governed by either roller speed, or the width of roller face open to pick up seed.

These drills normally have a hopper along the full sowing width, with one metering unit per row coulter. The row width for cereals is normally 110 mm or 190 mm. Wider placed rows for vegetables can be achieved by blocking intermediate outlets, but this will affect the calibration (see section 3.1.4 on calibration).

3.1.2.2 Purpose-built vegetable seeders

Specialist thin line vegetable seeders are produced with a variety of metering mechanisms. Most are made as single row 'units', so enabling a unit to be used in a hand-pushed frame, or in multiple units on a tractor drawn frame, which allows row widths to be varied.

The methods for metering seed from the hopper are as follows.

Brushing out of a hole. The discharge hole is towards the bottom of one side, and the brush is carried on a horizontal shaft inside the hopper, driven from the ground wheel. The speed of the brush, combined with hole size, dictates the sowing rate. Often a sliding plate with a selection of hole sizes covers the discharge hole.

Rotary plate. The seeds are carried out lodged in holes on a horizontal rotating disc; when out of the hopper the seeds fall through the disc hole into the coulter. Sowing rate is varied by disc speed, hole size and number.

Vertical disc. The seeds are carried out by cups, vanes, or holes in the edge of a disc rotating in the hopper bottom. Seed rate is varied by altering the speed of the disc, and the size and number of carrying points. On some cup drills the rate can be varied by changing the cups only, and omitting some altogether.

Gripper finger. Used for large seeds like sweetcorn; a number of cam-operated fingers on a vertical disc pick seeds from the hopper.

The plate, vertical disc and gripper types can sometimes plant seed singly, with a degree of spacing. Singulation is determined by the precision with which a seed fits the mechanism. Spacing is governed by the point of ejection; to be good, it must be at one set point, and close to the ground.

Centrifugal drills. A horizontal spinning disc is mounted in a chamber with exits to the coulter tubes around the wall. Seeds are fed on to the disc and thrown into the coulter tubes by centrifugal action. The disc speed and feed position are designed to distribute the seed flow evenly among the outlets. The disc unit requires little power, and is driven by a ground wheel. The coulters are fitted to a frame beneath the disc unit, which has to be mounted high enough to allow gravity flow. The production version has 24 outlets; to work properly, all have to be used. For drilling fewer rows, more than one outlet is placed into a coulter (with appropriate calibration), or surplus tubes discharge into a bucket.

The Snild drill is a small hand-pushed unit designed for transplant seedbeds. It consists of a small hopper, 300–500 mm long, with a shaft along the bottom, directly driven by a ground wheel at each end. The shaft has a series of dimples drilled in, positioned to drop seeds evenly, 25 mm apart in rows 35–50 mm apart. Simple openers beneath the hopper make the seed furrows, and small brushes on the hopper side restrict seed flow to one or two per dimple. The entire unit is no larger than a broom head, and is normally mounted on a brush handle. It is able to operate on a few grammes of seed.

3.1.3 'Precision' or spacing drills

These place seeds in pre-determined positions, and have enabled many crops to be drilled at the desired spacing, rather than in thin lines for subsequent hand singling.

Most drill mechanisms can perform correctly only on seeds that have regular shape and size, for example, spherical cabbage and turnip seed. Those of carrot and parsnip, for example, are not regularly shaped, so do not fit easily into the holes or cells of the singulating mechanism; other seeds are too small. To overcome these disadvantages the seeds can be 'pelleted'. This entails placing single seeds inside a small sphere of inert material, 2–4 mm in diameter. The material is designed to dissolve when in contact with soil moisture, and can incorporate pesticide chemicals if appropriate. Some seeds, such as onions, are regular in dimension and size, but angular rather than smooth. These are often formed into 'mini-pellets' with just enough filler to smooth the shape.

3.1.3.1 Types of precision drill mechanism

(a) *Cell wheel.* This is based on an aluminium wheel, about 150 mm diameter and 25 mm wide, which has a series of holes drilled around the edge. The hole dimensions allow only one seed to fit. The top of the wheel runs through the bottom of the hopper, where seed falls into the holes, surplus being held back by a wheel running in the opposite direction, the 'repeller wheel'. The seeds are held in the cell wheel until they are near the bottom of its travel, where they fall out into the furrow. To aid ejection, a thin groove is cut into the wheel edge, through the centres of all the holes, and a thin strip of metal running in the groove (the ejector) forces the seed from its hole. Normally a cell wheel is suited to one particular seed or pellet size, although the drill is designed to allow the wheel to be changed easily.

(b) *Belt feed.* Instead of a cell wheel, this uses a thin rubber belt with holes punched to suit the diameter of a seed; at the filling point the belt runs above a grooved baseplate, which determines the depth at which seed can settle in the hole. As with the cell wheel, surplus seed is rejected by a repeller wheel. By controlling projection of the seed above the belt, and the shape of the 'choke plate' which controls the feed from the hopper, one seed is held in each hole. The baseplate holds the seed in its hole until the sowing point is reached. Ejection is ensured by passing the belt round a small roller, where the hole widens by flexing. A set combination of hole size, baseplate and choke is needed for each seed or pellet size.

(c) *Cup.* A rotating disc, fitted with small cups, picks up seeds from the hopper, and drops them into the opened furrow. Unlike the system described earlier, the cup unit is at ground level and is designed to drop seed at a precise point during its rotation: the cups are free to rotate in the disc hole, and located by a projection from the rear of the cup stem engaging in a grooved track behind the disc. The track profile keeps the cup horizontal until it reaches the top of the disc, where it tips. The cup size and shape allow only one seed to be carried. Seed size variation is catered for by changing the cup units, rather than the whole disc assembly.

(d) *Vacuum method.* This uses air pressure to hold a seed against an orifice. The orifice has a special shape and size to ensure that only one seed can be held, although one orifice size can often cover a range of seed sizes; thus only three or four discs might be needed to cope with all seeds from cabbages to beans.

The orifices are drilled through a vertically rotating disc. At the point where each travels through the hopper it is subjected to vacuum conditions, so that the seeds are picked up. A series of carefully placed fingers knock surplus seeds away from the orifice. On some drills, ejection occurs simply as the disc emerges from the vacuum stage. This can be an erratic process and spacing can become unpredictable. These disadvantages are overcome by forming a cell around the orifice, so that the vacuum can be taken away prior to the drop point, and the seed is then held freely in the cell until it is required to drop.

While smooth, regular seed shape is ideal, most vacuum drills will work with any irregular shape that will block the orifice sufficiently to create the vacuum.

Drilling speed can affect performance; the simple mechanisms (belt, cell wheel and cup) are best below 5 km/h ($3\frac{1}{2}$ mph). Above this speed their accuracy falls because

(a) The speed of the mechanism, and the hopper agitation caused, do not allow the seed to fill every hole. This is not a problem with vacuum drills as the seeds are drawn towards the holes.
(b) The forward motion causes the seeds to roll along the furrow after dropping.

A drill metering mechanism developed by the N.I.A.E. overcomes the seed singulation problem by using two cell wheels, the upper one rotating slowly and acting as a feed to the rapidly rotating lower one. The rapid rotation of the lower wheel ejects the seed at high velocity so that it impinges in the furrow bottom to prevent roll. Roll can also be prevented by shaping the furrow opener so that soil is falling back to trap the seed as it lands.

3.1.3.2 Other drill components and assemblies

(i) All the above mentioned systems are made as single row 'units'; each is mounted in a chassis consisting of furrow opener, furrow closer and presswheel, and travels on a wheel in front of the opener and the presswheel. The metering cell wheel or belt requires little power, and can be driven from the front support wheel, which enables a single chassis unit to be fitted with a push handle to form a pedestrian-powered drill for small areas.

For large areas the required number of units are fitted to a tractor linkage toolbar, with a common drive shaft from a ground wheel powering all the units. The units are attached to the bar by a free linkage which allows each one to work at its correct depth. Some bars can apply a variable pressure to each unit by means of springs, to aid penetration in firm soils. Pneumatic drills require an air pump in addition to the disc drive; this normally confines their use to tractor-powered operation, but some battery-powered pedestrian machines have been made.

(ii) Some crops, like carrots, can be grown in double or triple rows, 15–20 mm apart. It is possible to achieve this from one unit if the cell wheel or belt has two or three rows of metering holes. Where the inter-row widths are slightly greater than the face width of the belt or cell wheel, angled plates beneath the drop point are used to deflect the seed outwards.

The width of a unit is 150–200 mm which determines the minimum single row spacing that can be sown by units mounted side by side on a toolbar. Where closer spacing is required, a 'tandem bar' system is adopted. This is mounted parallel to the front bar, about 1 m behind, and allows further units to run behind those on the front bar, drilling into the gaps between them.

(iii) The soil-engaging capability of the drill unit is an important factor in good establishment. Ideally, the seed should be in firm contact with damp soil, and have sufficient soil above to prevent it from drying out and to support it during germination, without imposing restrictions on growth. The major factors in optimum drilling and covering depth are soil type and weather.

In dry weather the damp soil is found at a greater depth, and the covering soil has to be pressed firmly to hold the plant as it germinates. On most drill chassis the furrow bottom depth can be varied in relation to the support wheels, and the degree of compaction above the seed controlled by the presswheels used (figure 3.1).

The N.I.A.E. has developed a drill chassis that is better able to deal with changing soil conditions. In this a small plough ahead of the furrow opener moves any dry or cloddy surface soil aside to expose the damp tilth; the seed is drilled into this and pressed into the furrow bottom before covering. The amount of dry soil returned over the seed can be varied, as can the degree of final pressure (figure 3.2).

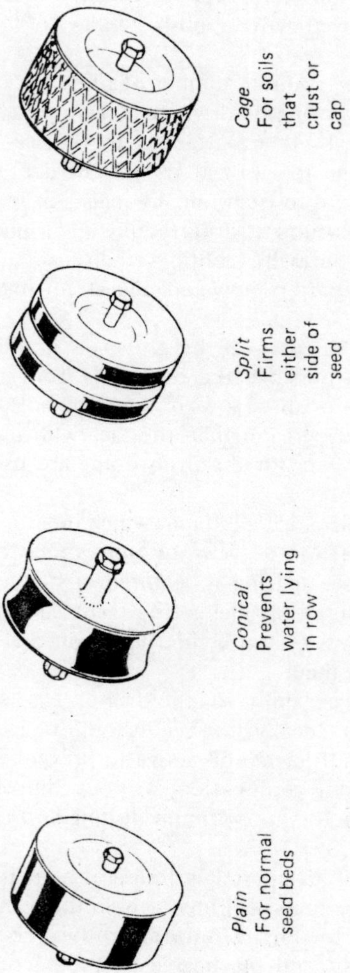

Figure 3.1 Presswheels.

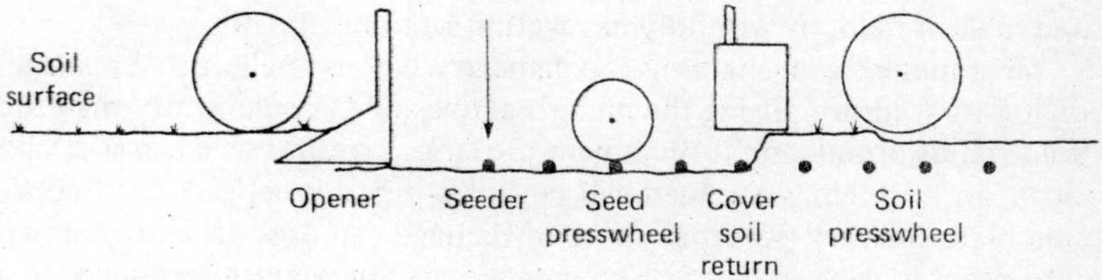

Figure 3.2 N.I.A.E. drill technique.

(iv) Monitoring systems can be fitted to warn the driver of malfunction. Three methods are used

An electric contact fitted to the metering mechanism switches a light on and off as it rotates. The lights are mounted in front of the driver. If the mechanism stops, the light remains on or off, instead of rhythmically winking.
A sensor in the hopper warns the driver when seed is running low.
A photo-electric sensing system detects seeds dropping from the metering mechanism. The rate at which the seeds should fall is preset, and if the required number of seeds has not fallen in the time set, the driver is alerted.

3.1.4 Seeder calibration and setting

The calculation of sowing rates is described at the end of this chapter (section 3.11).

Seed rate calibration of the simple broadcaster is done by measuring the distance required to spread a known weight of seed.

More sophisticated broadcasters, and thin line and precision drills, can be calibrated prior to operation by running only the metering mechanism and catching the seed delivered. The drill instruction book should give the method to be used, as well as the number of turns required to simulate the drilling of a known area. On most drills a handle is attached to the metering unit drive; where this is not possible the drill has to be jacked up and the drive wheel rotated.

The number of seeds discharged per metre of row by precision drill can be counted by the foregoing method, but not their spacing. Spacing action can be checked by running the unit over a clean, flat floor. A refinement of this is to mount the drill in a 'sticky conveyor belt' rig, which simulates travel in the field. The adhesive on the belt traps the seeds where they land, so the spacial pattern can be judged. This service is sometimes available from seed houses or drill suppliers.

Lateral spread pattern from broadcasters can be judged by laying out a series of shallow trays across the line of travel. After adjacent bouts have been spread, the quantity of seed in each tray can be assessed.

The drilling depth, covering and pressing has to be assessed each day, and in each field, by carefully excavating sections of row.

Most metering mechanisms can damage seeds and pellets by breakage, owing to seed not fitting the metering holes; if the hole is too small the seed stands proud, and if the hole is too large a second seed can partially sit in. In both cases the seed will be cut as it passes beneath the hopper side plate (choke). Another cause of damage can arise if seed is drawn into other moving parts. If this happens, clearances must be reduced to a point where the seed cannot enter – not increased until the seed can pass without damage. A sample of the actual seed used should be run through and analysed prior to drilling.

Chemical dressings and other substances on the surface of the seeds change their flow characteristics. Thus it is essential to do the calibration on the actual seed lot to be used. Some products also build up and clog the metering mechanism; this is not confined to 'sticky' liquids, as excessive dry power can fill holes in cell wheels and belts.

Precision seeders' mechanisms can be affected by the small variations in seed diameter between different seed lots. Most seeds and pellets are graded into size bands denoted by letters (table 3.2).

Table 3.2

Graded seed size coding

Code letter	*mm*	*Code letter*	*mm*	*Code letter*	*mm*
E	1-1.25	J	2-2.25	N	3-3.25
F	1.25-1.5	K	2.25-2.5	P	3.25-3.5
G	1.5 -1.75	L	2.5-2.75	Q	3.5-3.75
H	1.75-2	M	2.75-3	R	3.75-4

Source: Elsoms seeds.

Standardisation between seed suppliers and drill manufacturers ensures that the seed grading letter agrees with the metering system hole. Sets of hand sieves can be obtained to enable poorly graded seed lots to be split into suitable grade bands for drilling.

3.2 TRANSPLATING WHOLE PLANTS

Plants for transplanting will be grown either directly in seedbed soil (bare root or peg plant) or in modules of peat or other compost.

3.2.1 Plant raising

3.2.1.1 Bare root systems and production

Establishment is done in small blocks with the plants set close together in narrowly spaced rows, and drilled with any of three types of frame drill. The simplest is the Snild, described earlier in section 3.1.2.2. Better spacing is obtained from a special cell wheel drill with a number of cell wheels mounted side by side on a shaft, and fed from a common hopper. It will drill 10 or more rows 35 mm apart simultaneously, even though it is hand propelled and still small enough to operate in frames. A small vacuum seeder has been developed for larger areas. Like the previously described drill, the vacuum plates are on a common shaft to achieve close spacing. This machine is pedestrian controlled, but requires a petrol engine to power the vacuum pump.

As it is possible to control the environment and soil conditions in a plant bed, the drills have only simple coulters. The seed needs only light covering, or might be left uncovered, so presswheels are unnecessary. The plant bed system often uses only small quantities of highly priced seed, so the frame drill has to operate correctly with a few grammes in the hopper.

3.2.1.2 Module system production

This can be divided into two main areas – moulded blocks and containerised modules.

Moulded 'peat blocks' are made from compressed peat based compost; they are normally cuboid and the sizes range from 20 to 60 mm side length.

A hand tool to produce these blocks in small quantities consists of a number of block-shaped cups on a handle. The cups are pushed into the peat heap to be filled, and the blocks ejected by operating a lever on the handle. The correct method of operation is to lift the tool off the formed blocks at the same time as the ejector is used, rather than holding it above the seed tray and allowing the blocks to drop out. For production of larger quantities, a blocking machine is used (figure 3.3).

The machine works by first compressing the peat into a layer, equal to the block height, and then cutting it into cubes with the die plate. The compression plate and die plate reciprocate vertically, and the conveyor drive is intermittent, so that the peat band is stationary while the die is cutting. Output can be from 3000 to 12 000 blocks/h on a small machine and 11 000 to 25 000/h on a large one, depending on the size of block. The rate at which the peat band is extruded normally does not vary, so fewer large blocks can be cut per hour. Block size is altered by adjusting (a) the feed gate and compression plate (height), (b) the number of cells in the die (width), and (c) the distance that the conveyor moves between die cuts (length).

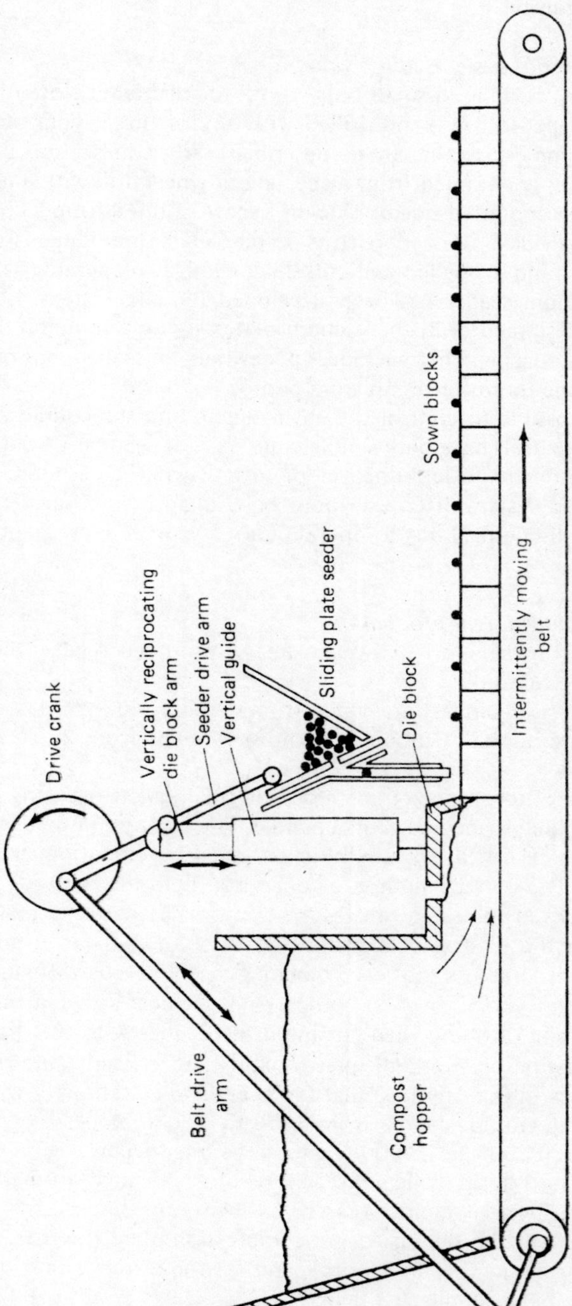

Figure 3.3 Blocking machine layout.

The blocks are taken off the machine by a special fork with closely spaced, straight, flat tines which slide under the blocks on the belt. They are unloaded on to the propagation area by gently pushing them off the tines with a lever-operated plate sliding from the handle end. Some very large blockers can feed directly into propagation trays or on to the floor. In the latter case the machine is on powered wheels which move it forwards in time with the die.

As peat is springy, the die cuts between blocks close back to a narrow crack. This can result in roots growing from one block to the next, making separation difficult. Various attempts have been made to produce discrete blocks, resembling the squares of a chocolate bar, using very thick (up to 20 mm) wedge-shaped cutters on the die block. The action of these thick cutters can compress the peat sideways and locally alter its moisture content, making it stick to the die plate.

Being only lightly compressed, the block is extremely fragile until it is bound together by plant roots. The blocks cannot be handled individually until this happens, and have to be transferred by the fork directly to the propagation area. As root binding has occurred by the stage where they require to be handled, artificial binding agents are unnecessary.

Containerised modules are becoming popular among vegetable growers, and have been used in the bedding plant trade for many years.

Vegetable cell modules are normally smaller than blocks, the current size being 17 mm square by 30 mm deep; one U.S. system uses cells that are only 10 mm in diameter by 25 mm deep. In most systems the modules are handled in their container or cell pack, so shape, size and stability of the module are no longer the limiting factors.

The most common container is the cell tray, made either of expanded polystyrene or blow moulded thin plastic. A system evolved in the Far East, commonly called 'Japanese paper pots', is a honeycomb of paper cells. Each cell is a complete paper 'tube', which is then glued together into a matrix. The glue is water-soluble, so that it handles as a honeycomb block before propagation, but disintegrates into individually wrapped cells for planting. All that is required for cell filling is to spread the compost over the cells, work it in and scrape off the surplus. Some long, thin (100 x 10 mm) Japanese paper pots are vibrated during filling to aid compost flow. Fully automatic cell tray lines are now used by large-scale propagators (figure 3.4).

The feed ram intermittently pushes the trays through in cell width steps, so that it is stationary for tamping and seeding. Compost is deposited on to the tray by the auger, and the surplus is ploughed back into the hopper. For this machine the compost has to be specially formulated and milled so that it will work its way into the cells by gravity. It must withstand 'unmixing' as it recirculates. Very long fibres have to be avoided as they can catch on the plough and be pulled out of the cell, bringing out the rest of the compost.

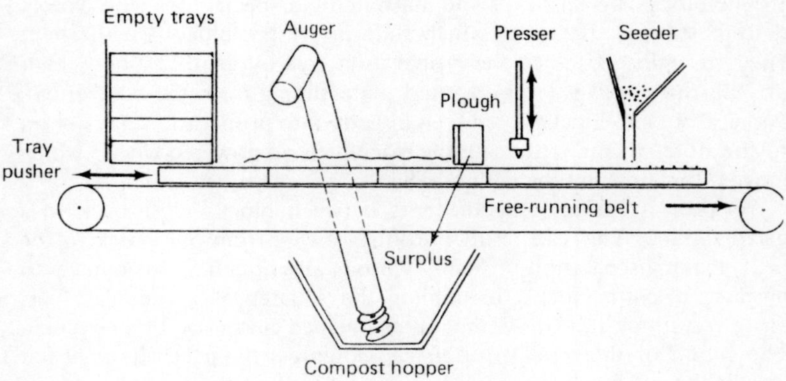

Figure 3.4 Module tray filler.

Paper pot fillers are slightly different in that the honeycomb has to be opened up and held until it is packed out with compost. The cells are inverted for filling, and sit on a plate with a small dome under each tube. When the matrix is turned over the domes have formed small depressions for seeding. The Japanese use soil in much of their propagation.

Attempts to develop a fully automatic transplanting system have resulted in many unique module systems. Some of these will be described later, together with the transplanter.

3.2.1.3 Seeding mechanisms for modules

Peat block modules can be seeded as they extrude from the blocker. The same system can be used on cell tray-filling machines, but other methods are available for seeding compost-filled trays in one operation.

Plate seeders are the most common type. A cross-section of one is shown in figure 3.5(a).

The reciprocating plate slides into the hopper where a seed (or seeds) lodge in the hole and, on return, surplus seeds run off or are dislodged by the brush. At the other end of its travel the seed is able to drop through into the planter tube. The hole diameter in the plate must correlate with the size of seed being sown. Where blocks have to have more than one seed (multi-seeding), the hole is bored to hold the desired number of seeds in a single layer. The number of seeds sown by this method can vary by one or two either side of the desired number, but the results are more consistent than those from a group of single seed holes.

The vacuum seeder has a bank of nozzles which reciprocate (figure 3.5(b)) to pick up seed from the tray and deposit it into the feed tube. As the nozzle tip enters the seed tray, suction is applied to pick up the

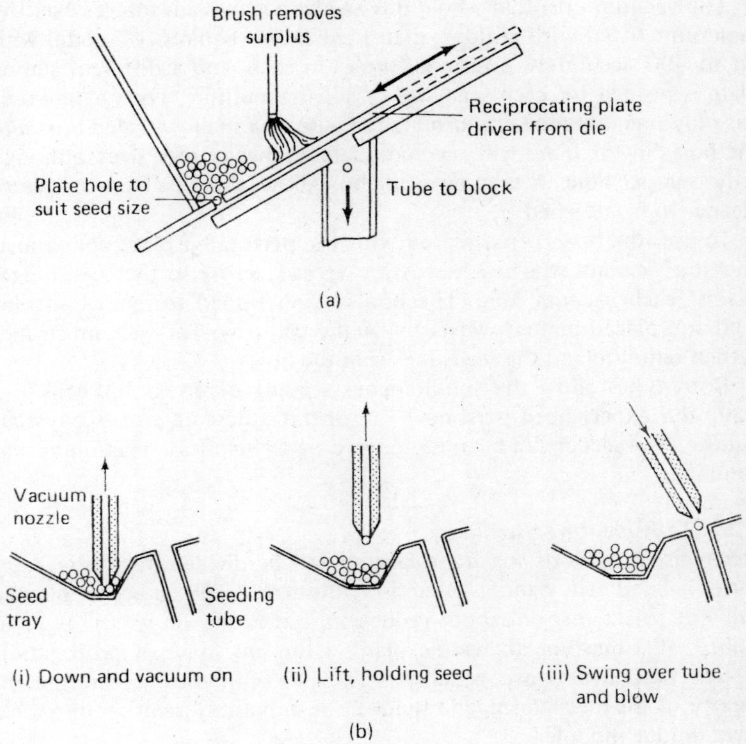

Figure 3.5 (a) Sliding plate seeder. (b) Vacuum module seeder.

seed. When it swings over the feed tube, the vacuum is released so that the seed can drop. The production machine has nine nozzles; each nozzle can have two holes if required, and therefore is capable of placing 18 seeds at once. The seeder is powered by compressed air, using air rams for movement, and a venturi ejector to create the vacuum. It is phased with the blocker by means of an air switch on the die plate.

The whole tray seeder with sliding plate consists of a shallow tray with holes in the bottom corresponding to the cells in the module tray. Covering the base is a sliding plate with a similar hole arrangement, these being bored to accept one (or the desired multiple) seed. This slides so that its holes are either in or out of phase with those in the tray bottom. In use, the plate is moved so that the holes are out of phase; seed is spread across and, when all holes are filled, the surplus is tipped to one end. The seeder tray is then placed over the cell tray, and the plate moved to bring the holes into phase so that the seeds can drop into the cells.

The vacuum operated whole tray seeder has two advantages over the same unit fitted with a sliding plate. The latter requires two plates with up to 400 accurately positioned holes in each, and a different sliding plate is needed for each seed size or sowing multiple. The vacuum unit has only one drilled plate, forming the base of a shallow sealed box, and the holes in the plate will accommodate a range of seed sizes, although only one per hole. A tapping in the box allows the pipe from a vacuum cleaner to be attached.

In use, the box is first placed with the perforated plate uppermost, and the vacuum attached. Seeds are spread across so that one lodges against each vacuum hole. The box is then tipped to remove surplus seed and placed plate downwards on the cell tray. The vacuum cleaner is then removed and the seeds fall from the holes.

Both types allow the simultaneous seeding of up to 400 cells in a tray, and experienced personnel can operate these at 3 or 4 trays per minute. The seeder can be mounted in a rig to help tray positioning and handling.

3.2.2 Transplanting machinery

Mechanised methods for transplanting can be divided into three categories – hand fed, hand assisted and automatic. The choice of planter will not solely rest on labour reduction, but rather on its ability. The ability of a machine to handle plants from any type of propagation system described above generally decreases with the increasing complexity of the mechanism, and the fully automatic systems require their own, unique module.

3.2.2.1 Hand feed

A 'hand-fed' planter is one that requires each plant to be handled up to the point of planting. The various machine designs are as follows.

(a) Peg roller – a drum with protrusions equal in size to a cuboid block. This is rolled across the soil surface to make indentations in which blocks or similar modules can be placed. The roller can be an attachment to a pedestrian tractor, or incorporated into an independently powered frame which also carries the planting gang.

(b) Furrow placement – the planter unit consists of a furrow opener and presswheel closer mounted on a frame which carries the operator who places plants in the opened furrow and holds them until the presswheels firm them in. Regular spacing is achieved by a ground-driven wheel, with a bell or clicker telling the operator when to place the plant.

A large version of this machine is made for planting nursery stock, and will handle whips over 1 m long. In order to accommodate this size of transplant, the operator normally sits to one side of the row. The large furrow required often limits the planter to a single row unit.

(c) Gripper wheel – as for (b) but fitted with a vertical gripper wheel 200–300 mm in diameter (figure 3.6(a)). This offers a more upright position, and pre-determined plant-placing positions. It also avoids the need to manually hold the plant until it is firmed in. Spacing is set by a combination of the gripper wheel driving ratio, and the number of gripper positions. One model grips the plant between a pair of thin springy discs which are forced apart at the loading and planting points. Plastic pegs on the discs locate the plant stem. Another version uses cam-operated gripper fingers on a single wheel, which can hold the plant by its stem or the module block. On both types the plants are placed in the wheel upside down (that is, with leaves towards the axle).
(d) Chain gripper – a further development of (c) where the plant gripper units are mounted on an endless chain (figure 3.6(b)). This enables the filling point to be raised to the best height for the operator, and two operators can sit either side of one chain to increase the planting speed.
(e) Powered dibber – used for planting leeks and other long stemmed crops, it has a vertically reciprocating head fitted with dibbing tines. To prevent soil adhering to the dib as it is pulled out of the hole, each one carries a large washer, backed by a spring, which slides down to hold the soil surface around the hole. To allow the tractor to continue to move forwards while the dib is in the ground, the head is free to move backwards against spring pressure. Plants are dropped in by a gang walking behind. Dib sizes can range from 25 to 50 mm. The soil has to be deeply cultivated to produce a good hole.

In terms of versatility, the pegged roller machine requires a block or module with a flat base, and cannot cope with bare root plants. The dibber will take bare plants and modules small enough to fall down the hole. The other types will take bare roots and most modules, but the block-gripping type requires a reasonably stable block.

The planting rate of hand-fed machines will be in the region of 1500–1800 plants per operator per hour.

3.2.2.2 Hand assisted

Hand-assisted planters allow the operator to feed plants randomly into a simple dispensing mechanism. They normally require modules to stabilise and separate the plants. The mechanism for dispensing plants falls into three types.

(a) Carousel – a horizontally rotating wheel with plant-holding cups around the edge. The operator has only to separate the modules and drop one into each cup, from which they are released down the coulter tube at the desired interval. Plant spacing is altered by the speed ratio between the carousel and the ground. As the plant has to drop 500–750 mm to the ground, the module is needed to provide weight

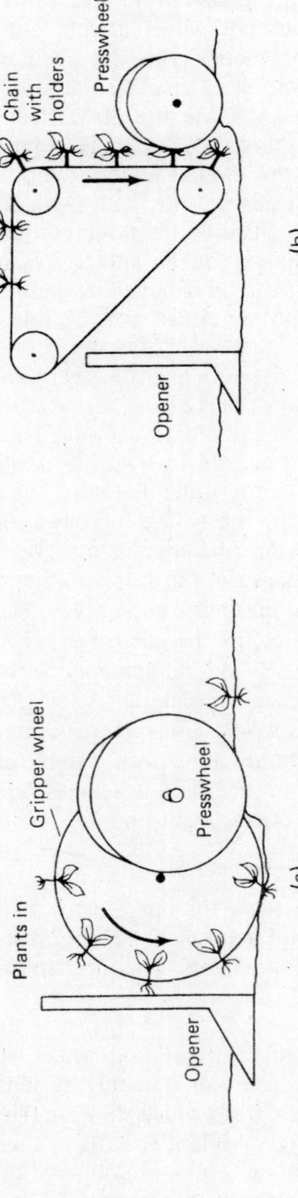

Figure 3.6 (a) Gripper wheel planter. (b) Carrier chain planter.

bias so that it falls root first. The cup shape and size are important in holding the plant upright, so that it will slide out without jamming. In some machines the diameter and length of cup can be changed to suit the plant size. The number of cups on the carousel varies between four and thirteen; initial observations suggest that workrate is not determined by the number of cups.

(b) Inclined belt – cuboid blocks are placed on to a conveyor belt which runs down to ground level and discharges them singly into the furrow. Rows of blocks can be taken out of the propagating tray and placed on the belt without further singling. On one type, the block discharge rate is governed by the speed of the belt, on another the belt is gravity-powered, and a reciprocating arm pushes off the lowermost block. By handling blocks in groups, the operator can theoretically service two or more rows, and achieve a high output per man. The module blocks must have a broad, flat base to prevent them tipping over on the belt, and cannot be cell tray grown if they are to be picked up in rows.

(c) Paper pot system – uses the 'paper pot' modules described in section 3.2.1.2. The outside row of tubes is picked off the honeycomb matrix with a special comb, and the tubes are laid, on their sides, on a conveyor with matching pockets which takes them down to the coulter and dispenses them singly. Like the inclined conveyor machine, one operator can theoretically service two or more rows. The output of these systems should be 2500–3500 plants per operator per hour.

3.2.2.3 Automatic

Many design concepts have been tried for a planting system that does not require any manual assistance in selecting or handling the plants. All the systems that have shown a chance of success have been based on special 'mechanisable modules', rather than on the automation of a manually handled transplant. This is illustrated by two examples.

(a) N.I.A.E. Bandolier. This is based on peat blocks, but initial research showed that ordinarily produced cuboid blocks could not be split or cut apart mechanically because of their dimensional instability, and the root growth between blocks. The N.I.A.E. system utilises a cylindrical block, with its sides wrapped in paper to contain the lateral root growth. The paper wrappings are joined into a chain or 'bandolier' of standard pitch, which enables the planter to locate the point where it is to cut them apart. The present bandolier system is standardised on a block 37 mm in diameter, and 37 mm high. The blocks are made by filling metal formers with dry compost and then injecting water in at high pressure. The wrap is formed from two strips of plasticised paper, heat-sealed together either side of the former. This is then withdrawn, leaving the block in its wrapping. The block and link dimensions allow the bandolier to be tightly nested in a honeycomb pattern for maximum propagating space utilisation.

The planter head pulls the bandolier string through a cutter to separate the blocks which then slide down the planting chute. Block spacing is governed by the rate at which the bandolier passes through the cutter head. The paper wrap prevents damage to the block and roots from the bandolier feed wheels and is designed to biodegrade in one or two months.

(b) The plug system is based on a special synthetic 'compost' formed into a small module, 10 mm diameter x 25 mm long. The material is extremely tough, while allowing root growth, and survives handling by the planter gripper mechanisms.

The plugs are purchased ready formed and are propagated in an expanded polystyrene cell tray. The planter mechanism pushes the plugs out of the cell holes sufficiently for them to be mechanically gripped and taken to the planting point.

In both systems the plant is handled by its module, thus preventing damage by the mechanism. However, both modules are specialised, the bandolier because of its wrapping, and the plug because of its composition. One disadvantage of handling by module is that it will be planted irrespective of whether or not it contains a plant; thus a propagation system is required to give 100 per cent viable plants. Other automatic designs have incorporated a stage where the plant itself is gripped, thus eliminating empty modules.

Planting rate is unknown at present but, when fully exploited, could begin to match the precision drilling rate. The planter will still require one operator to replenish the supply of plants.

Both these and other automatics will handle only the plant system designed for them but, if they break down, the special modules can still be planted by a hand-fed machine.

3.2.3 Transplanter operation

(a) The planter must leave the plant upright and at the correct depth. Some sophisticated machines are poor at this and require additional labour to walk behind the machine and make adjustments, as necessary.

(b) In outdoor situations, blocks and modules must be buried to prevent their rapid dehydration. This normally precludes pegged roller machines.

(c) The plant-handling system must match the planting rate to prevent time wastage.

(d) Some very small modules require water for establishment, either by irrigation or planter applied. Many planters have a water system that floods the furrow to water in the plant.

(e) Experimentation has shown that greater benefit is derived from planting thoroughly wetted blocks or modules, rather than from watering afterwards.

(f) An experienced operator on a hand-fed or hand-assisted planter will ensure that only sound viable plants are sown; this cannot be done with an automatic machine.

3.3 PLANTING CORMS, TUBERS ETC.

This category includes crops like potatoes, garlic, flower bulbs and onion setts. It is normally sufficient to scatter these in a furrow without attempting orientation, although some crops, such as potatoes, require regular spacing. While these products can appear dormant, in comparison to transplants, they are still susceptible to mechanical damage.

3.3.1 Types of planter

3.3.1.1. Hand feed

This can be similar to the transplanter in which each object is placed by hand in the furrow bottom. Where accurate spacing is desirable, machines can have a knobbed wheel running in the furrow bottom to leave indentations at the planting positions.

More sophisticated hand-fed machines have a vertical cup wheel. The operator drops a tuber into each cup, and the wheel speed is arranged to dispense them at the prescribed spacing. Some cup wheels are constructed so that the cups run slower and closer at the filling point to help the operator.

3.3.1.2 Hand assisted

One common type uses a conveyor belt with shallow cups moulded in to convey tubers to the planting furrow. One tuber is conveyed in each cup, and belt speed determines spacing. An operator is required only to ensure that each cup contains one tuber. The planting rate can be twice that of hand feeding.

3.3.1.3 Automatic-belt feed

The double belt machine has a pair of flat belts arranged to form a V-shaped trough. Both belts run in the same direction but at different speeds, which helps arrange the tubers into a single row, so that they drop off singly into the furrow. Tubers are metered from the hopper by a reciprocating trough and agitator. On some machines an operator controls the rate of tuber flow by adjusting agitation or hopper angle; the operator is able to control two planter units.

The multi-belt machine has a trough of 11 thin rubber belts, driven so that the five forming the base run towards the furrow, the rest on either side run in the opposite direction. This arrangement allows only a single row of tubers to reach the furrow, excess ones on either side of this line being carried back until they can roll into the centre. The operator stands above the planter mechanism and tips trays of tubers on to a flat belt which feeds the planter belts.

These planters are capable of operating at high speed – up to 7 km/h (5 mph). Their planting accuracy with potatoes depends on the size variation in the tubers but if graded in 5 mm increments the result will match those of other machines. They were designed for potatoes, with rates of 2.5–4 t/ha (1–1½ tons/acre), and rate is mechanically adjustable within these limits. Other crops can be planted if restrictors are fitted for reduced rates, or all flow impediments removed for high rates of up to 15 t/ha (6 t/acre).

Flat belt planters are also specially designed for the high rates associated with flower bulbs; these are discharged from the hopper into the furrow as a mass, with no attempt at singulation or spacing. The high planting rate requires a wide band, and the furrow opener can be up to 300 mm wide. Flow rate is regulated by a feed gate on the hopper discharge. As bulbs do not flow easily, powered agitation is fitted, and an operator normally rides on the machine to clear blockages.

Small models of this planter have been built for planting tulips, other small bulbs and garlic.

3.3.1.4 Automatic-cup feed

These machines have a series of cups on an endless belt, which pass vertically through the seed hopper and pick up one tuber each. The cup shape and size are designed so that the cup holds one tuber, any additional ones being dislodged by agitation or as the belt changes direction over the edge of the hopper. On some makes a manually charged carousel automatically drops tubers into empty cups. To avoid damaging the tubers in the hopper by dragging cups through, most planters feed only a thin layer to the cup pick-up position.

The planting rate on these machines is governed by the speed at which the row of cups can pick up tubers; on potatoes this equates to about 5 km/h (3 mph). Faster rates are achieved by having double or triple rows of cups to feed one furrow.

Accuracy is aided by reasonably close grading (10 mm incremental) and good shape for flow (short chits on potatoes).

3.3.1.5 Automatic with electronic control

This uses an inclined belt with six rows of shallow cups to convey tubers out of the hopper. The belt slope aids singulation by causing excess tubers to roll back. The belt moves in steps, one set of cups tipping at a time on to the planter belt, so it carries a single row of six tubers. When this has cleared the inclined belt end, the next six cups tip. A detector senses whether all six cups are full; if not, a separate make-up belt is activated. The planter rate is microcomputer-controlled from a setting of the seed spacing required, and measurement of the forward speed.

3.3.2 Net growing

Where bulbs and corms are grown in soil with clod problems, it is possible to plant them inside a tube of net to segregate them from the clod for lifting.

Small amounts of bulbs can be hand filled into short lengths of tubular net, and planted in an open furrow. On a field scale the net 'tube' is formed from two widths, one laid under the bulbs, the other over, these being held together at their edges by soil pressure. The modification to the planter is the addition of a pair of net-holding spools. One feeds into the furrow ahead of the bulb-dropping point, the other after this point but ahead of the coverers. The spools are lightly braked to ensure that the nets are laid straight and are taut.

3.4 POTTING AND TRAY FILLING

3.4.1 Potting machines

Most of this equipment mechanises only the compost-filling action, although a few complex machines place the plant also. The basic potting machine falls into one of two categories, depending on whether the plant is placed during or after filling.

3.4.1.1. Simultaneous planting

This machine provides a stream of compost; the pot is held beneath this with one hand, and the plant is held in position with the other. The compost flow is provided from a conveyor coming out of a hopper; the hopper is normally beneath the potting bench, so that the compost falls back into it when no pot is in place to receive it. Larger machines have a multiple outlet conveyor to feed the compost, and a pick-up conveyor beneath the potting bench to return surplus. On some, the potting bench vibrates to aid compost packing.

On small machines the pots are hand-placed. Some larger units incorporate a rotary table to hold them, leaving the operator's hands free to place plants. Another operator handles the pots on and off this table.

Most simple potting machines can fill flexible containers, although most need both hands to hold them open.

3.4.1.2 Planting after filling

These machines automatically fill and firm the compost, and also bore a small hole for the operator to drop the plant in. The simpler machines have rotary tables, which allow one operator to place empty pots, and to plant and remove filled ones. High output potters have facilities to automatically destack columns of empty pots, leaving the operator only to plant and remove the stacks. Some have sufficient throughput to require the stream of filled pots to be divided on to two planting conveyors.

These machines will only take rigid or certain semi-rigid pots, and can require resetting if the pot size is changed.

The planter hole is bored by a small cone-shaped auger; dibbing is normally avoided as it compresses the hole sides. The auger leaves a small ring of bored-out compost around the hole, which is easily pushed back after planting. Compost moisture content is critical to auger performance; if it is too dry the hole collapses, if too wet it blocks the auger flight.

3.4.1.3 Automatic potting systems

Some systems have been developed for mechanically placing the plant in the pot. As with automatic transplanting, the plant has to be specially prepared for automation; none has yet been found that is capable of sorting and handling bare cuttings.

One system uses the cell tray technique, where the plants are propagated in modules. The mechanism pushes the module out of the cell so that its foliage can be gripped by fingers to transfer it to the pot hole.

The other system is suitable only for potting on, using perforated thumb pots that are planted complete with the plant. Clamps with programmed movement drop the thumb pots into the filled and drilled pots. The small pots have to be presented to the clamp unit on a grid tray, so that they are in positions that the transfer mechanism can locate.

3.4.1.4 Potting rates

The potting rate depends on the size of pots being filled. The output of a simple simultaneous machine might be

Pot size (mm)	*Output/hour/operator*
250	250
200	360
150	500
125	550

More sophisticated simultaneous machines can pot at up to 500 pots/h/operator on 250 mm pots or containers.

Post-filling machines with automatic pot feed can operate at 10 000 75 mm pots per hour.

3.4.2 Tray-filling machines

Potting machines with conveyor discharge can also deliver compost into trays, which can be hand-tamped or vibrated to settle the compost, and any surplus scraped into the hopper.

Simple automatic machinery has a hopper, beneath which trays can slide on a conveyor. The hopper bottom is open, and its side plates fit closely either side of the tray. The front and rear plates just clear the tray and hold back surplus compost. Trays are pushed through in a solid line, and each is filled level to the top. A magazine is fitted in order to dispense trays automatically.

The cell module filler described previously can be used for seed trays.

Many devices have been designed to mark lines or dib holes in filled trays to aid seeding or transplanting.

3.4.3 Pot handling

One constraint to high-speed potting machines is handling the filled pots. The most popular method is a hand-held fork where the tines locate under the pot rims, thus enabling five or more pots to be carried with one hand.

Where pots have to be rehandled many times, they can be placed in a rigid grid. Picking up the grid will lift the pots, but when it is put down the grid drops and allows each pot to seat on to the bench.

Container-grown nursery stock can be handled and stood on special pallets with socketed holes in their upper surface. The plant cane is pushed through the pot into a socket to prevent it blowing over.

3.5 CANE INSERTION

This is a laborious operation if done by hand. To avoid cane breakage, it should be gripped towards the bottom end which forms an uncomfortable working position. A hand-held 'cane planter' has been developed, consisting of a steel handle fitted with gripping jaws, with a foot plate at its bottom end. The cane is pushed into the jaws and set in by foot pressure. The jaws have a cam action, which allows them to be detached easily from the cane after planting.

3.6 TYING PLANTS TO SUPPORTS

Plants can be tied to canes or trellis with a hand-held tool. In this system the 'tie' is a length of tape, joined by a staple. The taping tool has a pair of large jaws which go round the stem and cane; they hold a run of tape, which is wrapped around as the jaws close, and joined by a staple gun at the end of one jaw.

3.7 LIQUID-BASED SEEDING

3.7.1 Fluid drilling

This technique involves pre-chitting seed and sowing it with 1–2 mm of radicle projecting. The various steps involved are discussed below.

3.7.1.1. Germination

This is done in aerated warm water, sometimes aided by artificial light. The germinator is a large vessel fitted with a thermostatically controlled heater and an air compressor.

3.7.1.2 Drilling gel

The seeds are field-drilled while suspended in a viscous gel, similar to wallpaper paste, which is obtained by mixing a powder with water. Some powders can be tipped into the water and stirred, others require to be mixed intimately with the water as they are run into the tank, which requires a venturi, where the powder is drawn into the water stream by the vacuum that it creates.

The chitted seeds are mixed into the gel with low-speed paddles to avoid damaging the radicles.

3.7.1.3 Block and tray sowing

This unit uses vacuum probes to pick up seed suspended in an aerated water trough. The probe bank then moves over the tray and the seeds are discharged by air pressure.

The reciprocating action is timed to coincide with the passage of the tray or block beneath.

3.7.1.4 Field sowing

Several systems have been tried to extrude a uniform line of gel into the furrow.

The simplest method is to force the gel from a sealed vessel by air pressure. The gel and seed do not have to pass through any moving parts, so there is little seed damage, but gel extrusion control proves difficult.

Centrifugal sludge pumps are simple, but the seed could be damaged by the turbulence created by the rotor.

Peristaltic pumps rely on rollers to squeeze the gel along a plastic tube. The rolling action results in intermittent flow which is smoothed by an air diaphragm vessel in the flow pipe. The system appears to do little damage and, being positive displacement, the flow rate can be reliably set.

As the seeds are randomly dispersed in the gel, they are not extruded at a regular spacing. None of the above systems can be regarded as a 'precision' method.

A chassis unit, similar to that of a precision drill, is used to form a normal furrow to receive the gel.

3.7.2 Hydraulic seeding

This is an entirely different process from fluid drilling, being a method for broadcasting seed on to areas that are inaccessible to vehicles. The seed is mixed with water, fertiliser and cellulose mulch. The resulting slurry is sprayed on to the seeding site using powerful jets. The equipment consists of a mixing tank, subjected to agitation to keep the materials in suspension, a pump capable of passing the slurry, and a nozzle gun. The gun can project the slurry to a distance of 60 m or more. The discharge rate enables large areas to be covered quickly: 0.1 ha in 6–7 minutes.

3.8 SEED PRIMING

This is a technique in which the seed commences its germination process, but is prevented from sending out the radicle. After priming, the seed is dried, and can be sown with a normal drill.

The process is carried out in an aqueous solution containing special salts which allow the seed to draw in sufficient water to metabolise, but not to germinate fully. The aqueous solution has to be aerated to maintain oxygen levels, and the solution strength must be accurately maintained. Some seeds, such as lettuce and freesia, can be treated in ordinary warm water, as the heat inhibits the final germination.

3.9 GRAFTING AND BUDDING

Although this is regarded as a manual operation, mechanisation can be used to improve certain aspects.

3.9.1 Rootstock preparation

Rose and similar rootstocks can be bared for budding by brushing or blowing soil away from the stem. Either machine can be mounted on a small tractor.

3.9.2 Budding

A hand tool has been developed to implant a prepared bud into the stock for rose culture. The tool resembles a gun; the sharpened end of the barrel tube is pushed against the scion, around the bud eye, and cuts it out, together with a small circle of bark. The gun is then held against the stock and the trigger is operated. The initial trigger movement causes a pair of blades to move out from around the barrel nose to the stock bark; the blades then move apart to open up two small flaps of bark. The final trigger movement operates an ejector which pushes the bud from the barrel tube against the bared stem.

This tool has been developed to replace the traditional knife budding technique.

3.9.3 Graft preparation

Where the scion is to be placed directly on to the end of the cut root-stock stem, a hand tool is available to cut and prepare the surfaces. This consists of a specially shaped blade which slices the stem with a V-shaped cut. The tool centralises the stem before cutting to ensure that the V is in the correct position. Both stems are cut with the tool, and the male V of the scion fits tightly into the female V of the stock.

3.10 PLASTIC-FILM MULCHING

This technique involves laying thin Polythene film over a seedbed to form a cloche. The film is obtained in rolls, wide enough to cover a 1.8 m bed. The laying machine has to implant both edges to keep the film taut and secure. Two implanting methods are used.

3.10.1 Ploughing

A small plough excavates a furrow, throwing the soil outwards. The film is pressed into this furrow by a rubber wheel, and a second plough body returns the furrow slice to trap it.

3.10.2 Disc pressure

The sheet is laid across the bed, and a blunt disc running about 100 mm in from each edge presses it into the soil.

Both systems have their advantages and problems. These can be summed up as follows.

(a) Soil conditions – the discs require soft, cohesive soil; the plough works best in dry soil.
(b) Working width – the plough type needs a wide space between edges of adjacent film runs.

(c) Film elasticity – the disc types stretch the film laterally during implanting. Slit perforated films accommodate this, others require to be laid with a degree of lateral slackness before implanting.

Lateral slackness can be put in during laying, by running the centre of the film over a raised plate or roller, while the edges are being implanted. Longitudinal tension can be varied by adjusting the speed ratio between the film feed rolls and the forward motion.

Most films have to be removed before they restrict crop growth. Power driven re-rollers are available but they do not normally produce a neat, tight roll that can be re-used. Some crops, such as sweetcorn, are grown through the film, this is achieved either by using a drill or planter, which perforates a solid (unperforated) film, or by slitting a hole above each emerging plant.

3.11 CALIBRATION THEORY

The following formulae can be used to arrive at the calibration rate for any operation.

(A) Distance of travel for an area

$$\text{metres/ha} = \frac{10\,000}{\text{working width (metres)}}$$

Notes. (i) Working width for spinning disc broadcasters = width between tractor wheelings from consecutive runs.
(ii) Working width for row crops = row spacing × the number of rows being worked.

For example, a 4 m wide drill will travel 10 000/4 = 2500 m to cover 1 ha.

(B) Weight dispensed in terms of distance travelled

$$\text{kg/metre} = \frac{\text{kg/ha} \times \text{width m}}{10\,000}$$

For example, for grass seed at 125 kg/ha, sown with a 3 m machine

$$\text{kg/metre} = \frac{125 \times 3}{10\,000} = 0.0375 \text{ kg/m or } 37\tfrac{1}{2} \text{ grammes/metre of travel}$$

Therefore, 1 kg of grass seed should run for 26.7 metres.

(C) The effect of using part of the full width

When using a drill with some coulters blocked to increase plant row widths, but where the seeding rates used in the manufacturer's calibration tables are in terms of overall area, the drill setting must be increased by

$$\frac{\text{total no. of outlets}}{\text{no. of outlets used}}$$

For example, for beans to be drilled at 145 kg/ha, in 4 rows at 440 mm row centres, with an 18 row drill having coulters at 110 mm centres, the drill will need to use only four of its eighteen coulters. Therefore

$$\text{drill setting rate} = 145 \text{ kg/ha} \times \frac{18}{4} = 652.5 \text{ kg/ha}$$

(D) Plant stand and spacing derived from plant population and row widths

(i) For equally spaced single rows

$$\text{plants/metre of row} = \frac{\text{row width (mm)} \times \text{plant pop./ha}}{10\,000\,000}$$

For example, for onions, 860 000/ha in 300 mm width rows

$$\text{plants/metre} = \frac{300 \times 860\,000}{10\,000\,000} = 25.8$$

$$\text{plant spacing} = \frac{1}{\text{plants/metre}} = \frac{1}{25.8} = 0.038 \text{ m or } 38.7 \text{ mm}$$

(ii) For beds – plant population is per overall field area, but only bed strips are actually planted, so

$$\text{plants/metre of row} = \frac{\text{bed wheeling centres (m)} \times \text{plant pop./ha}}{10000 \times \text{no. of rows/bed}}$$

If the onions were drilled in a 1.8 m bed at 5 rows/bed, then

$$\text{plant/metre} = \frac{1.8 \times 860\,000}{10\,000 \times 5} = 30.96$$

$$\text{plant spacing} = \frac{1}{30.96} = 0.032 \text{ m or } 32 \text{ mm}$$

(iii) The above are for *single* rows only. With twin or triple row techniques the plant stand/unit of row is the same, but the spacing between individuals in each row will be increased by two or three respectively. For example, if the bed-sown onions were sown in twin rows, the stand would still be 30.96 plants/metre, but they would be spaced at

$$\frac{1}{30.96} \times 2 = 64 \text{ mm}$$

in each row.

(E) *The above calculation supposes that all the plants are viable. When raising from seed the sowing rate must include factors for germination and field conditions*

Therefore, if the onion seed in (D) (iii) had a germination of 85 per cent and the field conditions allowed only 90 per cent of the germinated seed to grow, the drilled plant stand would be set for

$$\text{required stand} \times \frac{100}{\text{germination \%}} \times \frac{100}{\text{field factor \%}}$$

$$= 30.96 \times \frac{100}{85} \times \frac{100}{90} = 40.5 \text{ plants/m}$$

and the spacing in twin rows would become

$$\frac{1}{40.5} \times 2 = 49 \text{ mm}$$

Note that, if it is imperative to maintain a desired spacing, for example in drilled cauliflowers, a full stand can be ensured by dropping two or three seeds at each 'station', and removing the surplus plants later.

(F) *The weight of seed required can be calculated from plant population if the number of seeds per gramme or kilogramme is known*

$$\text{theoretical seed weight (kg/ha)} = \frac{\text{plant population/ha}}{\text{seed count/kg}}$$

$$\text{actual seed weight} = \text{theoretical seed rate} \times \frac{100}{\text{germination}} \times \frac{100}{\text{field factor}}$$

For example, the onion seed in (D) has a count of 2800/10 grammes. This equals 280 000/kg.

$$\text{actual seed weight/ha} = \frac{860\,000}{280\,000} \times \frac{100}{85} \times \frac{100}{90} = 4 \text{ kg/ha}$$

The forgoing calculations must be regarded as providing only the basis for setting a piece of application machinery, the full setting can be achieved only by referring to the manufacturer's handbook.

4 FERTILISER AND CHEMICAL APPLICATION

4.1 FERTILISERS

Most fertiliser for land application is purchased as pre-mixed compounds, either in solid or liquid form.

4.1.1 Solid fertilisers

Compound fertilisers come in one of the following forms:

(a) Granules. The mixture is reformed into granules of reasonably regular size, but roughly shaped.
(b) Prills. The reforming process results in small uniformly shaped 'beads'.
(c) Raw mixtures. The mixture is left in its raw state, containing everything from fine powder to coarse crystals.

The above forms affect the performance of the spreader in both rate and evenness of application. Prills flow more readily than granules, while raw mixtures can segregate in the metering mechanism or during spreading.

Machinery for spreading can be divided into three main types, full width, pneumatic and centrifugal.

4.1.1.1 Full width broadcasters

These machines have a long, narrow hopper, mounted at right angles to their direction of travel. The fertiliser is metered out of the bottom, directly on to the soil beneath. Various metering mechanisms have been used, the most common being the following.

(a) Single roller. The roller forms the base of the hopper. There is a small gap between the hopper side and the roller so that, when it rotates, it carries out a thin layer of material. The roller can be smooth, but often has small longitudinal flutes to grip the fertiliser (figure 4.1(a)).

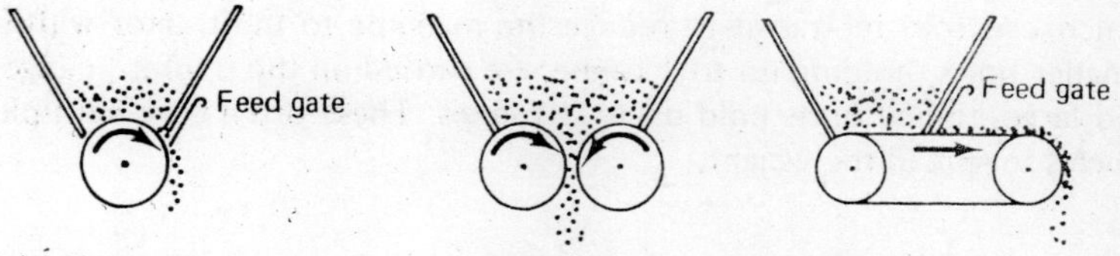

Figure 4.1 Fertiliser dispensing mechanisms.

(b) Twin roller. This is similar to the single roller except that the hopper base is formed by two contra-rotating rollers. They have a small clearance between them, and the material is pulled through this gap as they rotate (figure 4.1(b)).
(c) Belt. The single roller is replaced by a short, wide belt (figure 4.1 (c)).

All of these machines can be obtained in several sizes, from hand-pushed 'garden' spreaders about 300 mm wide, to 'field' spreaders 3-3.5 m wide. On large machines the width is normally limited by stiffness of the roller; over a certain length the roller can bend sufficiently in the middle to given uneven application. On garden models the roller is driven directly by the land wheels, and flow rate is varied by the aperture between the roller and the hopper side. Larger models combine a variable ratio geared drive to the roller, with adjustable apertures to give a wide rate capability.

(d) Hole and agitator. The hopper base has a series of adjustable apertures. A shaft in the base of the hopper carries agitator fingers which encourage flow from the aperture. The flow rate is adjusted by aperture size, but is very dependent on material flow characteristics.

4.1.1.2 Pneumatic

These machines use a series of air pipes to distribute fertiliser from a central hopper. The correct amount of material is fed into the air-stream of each tube by a force-feed type metering unit under the hopper. The metering unit is driven from the PTO in some mounted machines, or by a ground wheel in other mounted and in trailed machines; this latter method ensures that application rate is related to travel speed. It is possible also for the metering drive of a mounted spreader to be taken from the tractor's rear wheel. In all cases the application rate can be changed by a gearbox. Air supply to the distribution pipes comes from a PTO or engine-driven fan. Each tube carries fertiliser for a width of 600 mm, and is evenly spread within this width by blowing against a specially shaped deflector plate.

The pneumatic tubes are carried on a boom, normally 12 m wide, which can fold in transit to reduce the machine to the tractor width. Smaller units, holding up to 1 tonne, are carried on the tractor linkage, and large, trailed units hold up to 3 tonnes. These often have multiple wheels to spread the weight.

4.1.1.3 Centrifugal

These machines operate by throwing the material to either side of the hopper. Two systems are used for throwing, spinning disc and oscillating spout.

(a) The spinning disc consists of one disc or, on larger machines, two horizontal discs fitted with radial vanes on their upper surface. Fertiliser is metered on to the disc and thrown off by centrifugal action. On single disc machines the spread covers an arc of 180° to the rear; on twin discs, complementary arcs of 90° each are covered. These spread patterns are obtained from the point at which the material is fed on to the disc. Fertiliser can be thrown over 20 m each side, but the amount diminishes with distance from the disc. Uniform distribution is achieved by overlapping the next run to complement the distribution (figure 4.2).

The lack of sharply defined spread width can be an advantage when spreading areas without defined wheel guidance markings, such as grassland, as slight variation in spreading width will not be significant.

The disc can be driven directly by the ground wheels on small machines, by PTO on tractor mounted models, and by PTO or hydraulic motor on large trailed machines. Disc speed on tractor-driven units can be varied by PTO speed or oil flow. Direct ground driven spreaders normally have the disc speed fixed to the forward speed.

Many sizes of machine are made, ranging from small amenity grass models for hand or lawn tractor pulling, through tractor-mounted models, to large trailed machines, holding over 2 tonnes of material.

The metering of material from machines with a conical hopper can be done using an agitator and adjustable gate on simple machines, or else some form of variable force feed. Trailed machines with V hoppers meter by a belt or a chain in the hopper bottom, with variable speed, via an adjustable aperture. Large belt discharge spreaders can tackle coarse, lumpy material such as slag, lawn sand or sugar factory lime.

(b) Oscillating spout machines achieve their 'throw' by accelerating it along a swinging tubular spout. The basic distribution pattern is similar to that from the spinning disc type, but the pattern is more uniform. It is less likely to be affected by variations in the fertiliser. Spread pattern is aided by a horizontal deflector strip across the spout end.

These machines are available in small, tractor-mounted or large trailed versions with metering similar to disc spreaders.

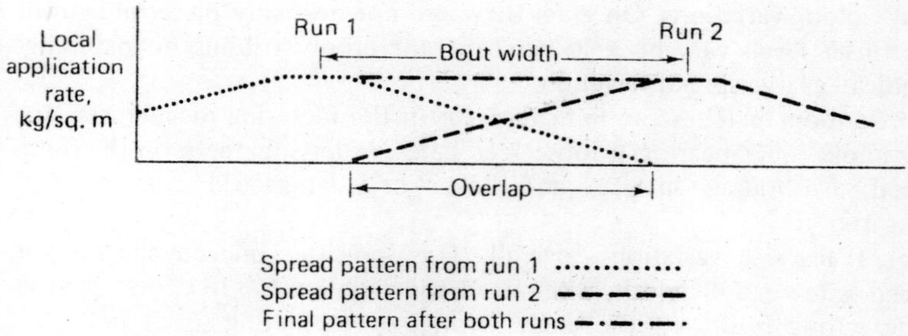

Figure 4.2 Centrifugal spreader pattern.

4.1.1.4 Calibration and spread accuracy

Calibration of a spreader that uses a positive metering mechanism can be done by manually cranking it for a known number of revolutions, which corresponds to an exact fraction of a hectare, and weighing the results. On simple machines it is necessary to operate over a calculated distance and measure the amount of material used. The amount can be gauged by filling the hopper level before the run, and seeing how many bags are needed to refill it to this level. The distance travelled to spread a set area can be calculated by the method given in section 3.11.

The transverse spread pattern should be checked regularly.

Lateral pattern is checked with a series of catching trays laid across the spreading width. The 'standard' tray is 1000 × 250 mm plan area × 150 mm deep. A honeycomb in the tray prevents granules from bouncing out. Trays are set out to form a line across the spread width, with gaps left where the wheels run to enable the spreader to drive across at its correct speed. After each run the volume of material in each tray is measured in a small measuring cylinder. If the volume from the trays is plotted on a graph, the spread pattern will emerge. A more visual record is obtained in a 'patternator', a number of glass or perspex tubes mounted together, each receiving the contents of one tray. The difference in column heights provides an immediate spread pattern. The tray system can be used in the yard or even a wide barn to provide a 'pre-season' check, but it should also be repeated in the field to assess the effects of ground conditions and wind. For spinning disc machines, the tray line must encompass the full spreading width. This run will show the declining application pattern. The final pattern will emerge if there is enough room available to carry out an adjacent run; if not, this can be calculated by adding the corresponding volume as it would occur from the adjacent run (figure 4.2).

Common faults with spread patterns are

(a) Colour variations. On grass these are not normally discernible until fertiliser levels vary by ± 15 per cent, therefore 'striping' or 'patching' indicates large spread variation.

Random patches can indicate wear in the metering mechanism (for example roller bearings), loose feed gates, or missing teeth on the force feed; also bounce on 'plate and flicker' or a damaged disc on a centrifugal.

(b) Transverse variation – overall. This indicates uneven aperture or feed gate on full width, disc speed too high or too low, disc feed in the wrong position, disc angle wrong, or oscillating spout deflector damaged.

(c) Transverse variation – stripes within spread width. This is caused by unevenly set or worn feed apertues, worn grooves on the roller or belt, a damaged plate on 'plate and flicker', damaged pneumatic deflectors, uneven pneumatic boom height, pneumatic pipe blockage, or a hole in the hopper.

4.1.1.5 Specialised application equipment

(a) Band application is practised in row crops, where overall application would be wasteful. The equipment normally consists of a set of individual hoppers with force-feed metering, positioned to drop material down tubes on to the row. Some pneumatic machines have facilities for blocking feed on individual metering rotors. The 'hole and agitator' hopper machine can be fitted with dropper pipes to direct the flow to rows. For widely spaced rows, two or three pipes can be directed on to one row.

Similar equipment is used on drills and planters for simultaneous nutrient application, to allow for 'placement' where fertiliser is deposited in the soil in a given position relative to the plant roots.

(b) Deep placement is becoming popular in certain crops, where nutrients are put at depths of 100 mm or more. This involves mounting an individual applicator on a subsoil tine cultivator, so that material is deposited into the soil as it heaves around the foot.

(c) Ridge placement is used for potatoes. One method is to ridge prior to planting and to scatter fertiliser on the ridge shoulder. When the planter splits this ridge over the tubers, the layer of fertiliser falls near them but is mixed into sufficient soil to prevent it from burning the emerging shoots.

4.1.1.6 Corrosion

Fertiliser chemicals readily attack the mild steel parts of spreaders so major components are now made of plastics or stainless steel, although many of the more important moving parts are still made of ordinary steels.

Metal wastage due to corrosion is normally unimportant. Of greater significance is that the products of corrosion occupy a greater volume than the parent metal and that the resulting 'swelling' can seize moving parts. These products can also flake off and block apertures.

Therefore, after use, all susceptible parts should be washed clean with water, and protected by a coating of rust-inhibiting oil.

4.1.1.7 Handling fertiliser

While the 50 kg bag is still the most common container for fertiliser, many large operators are receiving 'big bags'. These hold between 250 kg and 1 tonne, and are constructed like a large carrier bag, with handles that can be used for lifting with the help of a fork lift tine or crane hook. The bag is normally a strong woven outer with a waterproof liner, and designed to discharge through the base by a special hatch, or by cutting it open in the case of a 'one trip'.

The bag system represents an easy way of handling large quantities of granular material. It normally requires a forklift or heavy-duty tractor loader, although a few specialist spreaders incorporate a self-lifting device. The bag is normally fully emptied in one go, although the discharge neck can be temporarily tied up after partial emptying.

4.1.2 Liquid fertilisers

These come in three forms, solution, suspension and liequfied gas. The solution type is a free flowing liquid with all the nutrients dissolved in water. In 'suspension', there is insufficient water to dissolve all the nutrients, and they have to be kept in suspension by agitation.

Both types can be applied with a high output hydraulic crop sprayer. The application rate is 500–2000 litres/ha (50–100 gal/acre), depending on the crop. The sprayer is fitted with large-orifice, low-pressure jets; this not only gives a satisfactory application rate, but also produces large drops which roll off crops being top dressed, to prevent scorch. The liquid can also be applied in bands during the planting or drilling of vegetable crops, or injected into the soil under a growing crop by thin 'knife' tines which slit the soil, each tine carrying an injector at its lower end.

Liquid application rate is controlled in the same way as in hydraulic pesticide sprayers (see section 4.2.2).

One advantage of using liquid fertilisers is in handling, as the material can be transferred to the sprayer by pumping. The fully dissolved liquids are relatively bulky, and require large storage tanks. Suspension systems have reduced the bulk, but require specially agitated storage.

Liquefied gases are normally ammonia based. To retain their liquid form requires pressures of 550–700 kPa. At this pressure the liquid occupies 1/850 of its original gas volume. It is injected as a liquid into the soil, 100–150 mm deep, where it vaporises and combines with the soil water. The injection is normally done by cultivating tines. In loose soil these are designed to allow the soil to fall back into the slot to seal in the gas. In grassland, thin knife tines are used, together with rollers to press back the slot edges. One version uses hollow tined spiking wheels which release the gas when each spike is fully into the soil.

Liquefied gas systems require pressurised tanks at all stages of handling

4.2 LIQUID PESTICIDES

This is the most common form of applying pesticides, and usually involves creating droplets to coat the 'target'.

4.2.1 Spray droplet technology

One of the most important factors in efficient and effective spraying is the size of the droplet. Very small drops are easily carried away by wind (spray drift), and very large drops roll off leaf surfaces and are lost to the ground.

Drops are measured in terms of 'volume mean diameter' or VMD, the diameter that a spherical droplet would need to carry the same volume of liquid found in a real droplet. VMD is measured in microns, which are thousandths of a millimetre (table 4.1).

Table 4.1

Characteristics of droplet size

VMD (microns)	*Characteristics*
Under 50	Uncontrollable, rapid evaporation, will not target by gravity
50–100	Normally uncontrollable, still need assistance to impact on target
100–250	Less affected by wind, impact by gravity, good retention
Over 250	No wind drift, tend to run off plant surfaces

The second important consideration of drop size is the volume of liquid contained. The volume of a sphere is $\frac{4}{3}\pi r^3$, where r is the radius or half the diameter (table 4.2).

Modern spraying theory seeks to combine the factors contained in the two tables for maximum efficiency. Some systems intentionally produce extremely small drops – under 100 microns – and use other means of assisting the drops to their target. These can be fan blowing, electrostatic attraction or natural wind drift. Very large drops – over

Table 4.2

Droplet size to volume of liquid

VMD (*µm*)	*Radius* (*µm*)	*Volume* (*picolitre*)	*Drops/litre*
100	50	0.52	1923
150	75	1.76	570
200	100	4.16	238
250	125	8.13	123

300 micron – are used where leaf run off is essential, for instance when applying liquids to the soil beneath the crop.

The sector attracting most interest is a spray of droplets in the 100–200 micron range, which combines a high density of droplets with gravitational impingement and reduced tendency to wind drift. Agronomically, the interest is towards reducing the rate of both carrier liquid (dilution water) and chemical, which might be possible if the spray could be reliably formed into drops of this size. This has implications for reducing the soil damage caused by the sprayer weight and for making savings in chemical costs.

Modern testing equipment allows an analysis of the droplet size in the discharge from an atomiser unit to be made. Any spray will consist of a range of differently sized droplets. The atomiser characteristics can be seen by plotting each drop size band against its percentage of the total volume (figure 4.3).

All atomisers produce a droplet size distribution in the form of an 'S'-shaped line. The desirable shape of this line is for the centre to be as long and as near vertical as possible. The curved portion at the top represents the percentage of over-large drops, that at the bottom, the percentage of overfine drops.

Figure 4.3 also shows the effect of altering the operating pressure on a good nozzle. Raising the pressure moves the line to the left (that is, it produces finer drops), but the amount of undesirable sized drops does not change significantly. The same effect would accrue from raising the speed of a rotary atomiser.

4.2.2 Hydraulic spraying

4.2.2.1 Jets

The droplets are formed by forcing a liquid through a specially shaped jet. The pattern of droplets that emerges depends on the shape of the orifice, the three most common patterns being shown in figure 4.4.

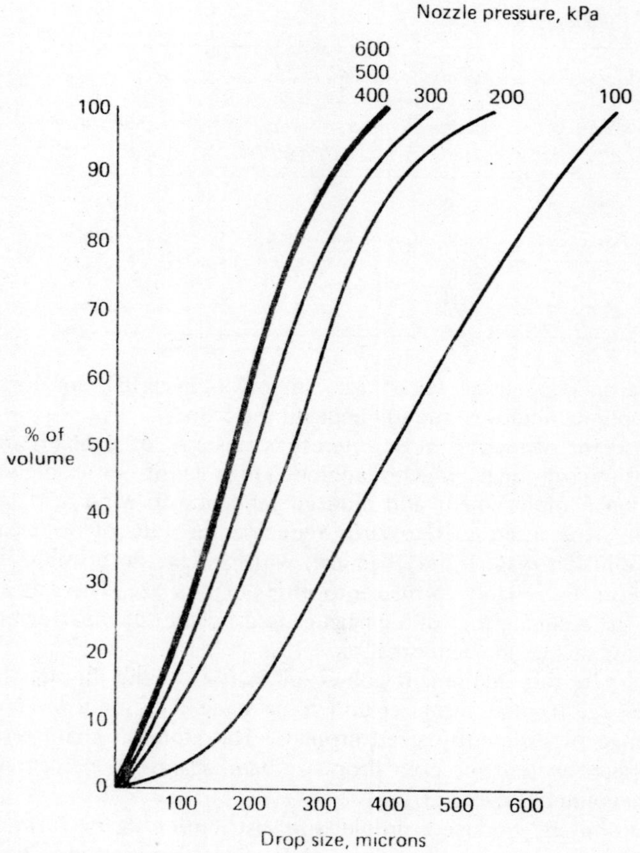

Figure 4.3 Droplet spectrum of a nozzle.

The fan jet (figure 4.4(a)) is ideally suited to a multi-jet boom, because its distribution is complemented by the overlap from those on either side. Its pattern is produced by a single small slit.

The cone types, having a more definite cut-off, are well suited for band spraying and hand lances. The most common cone jet is made of two discs (figure 4.4(b)) spaced about 3 mm apart; the inner one (swirl plate) has a number of small louvred slots which cause the liquid to spin as it flows through. The outer disc is convex and has a single central hole. The hollow cone is produced by a solid centred swirl plate, the solid cone by one with a central hole. Cone patterns also can be produced by nozzles with a helically grooved swirl plate or by feeding in the liquid into the swirl chamber tangentially.

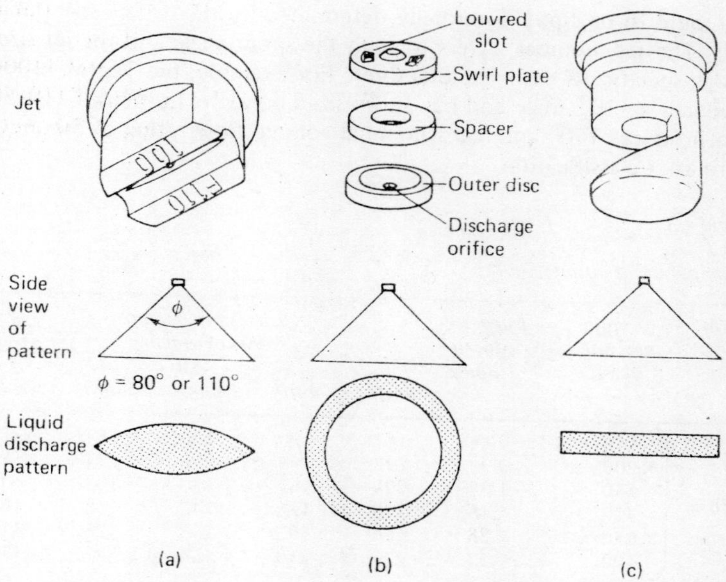

Figure 4.4 Jet spray patterns: (a) fan, (b) hollow cone, (c) anvil (flood jet).

In the anvil jet (figure 4.4(c)) a stream of liquid issuing from a plain bored hole is deflected by impingement on an angled 'anvil'. The angle and shape of the anvil produces a 75–90° arc of spray, with a uniform distribution.

Jets must withstand corrosion and erosion by the chemicals used. Normal materials used are brass, plastics, ceramics and stainless steel; brass and plastics can suffer from rapid wear but their need for frequent replacement has to be set against their lower cost.

The spray discharge angle in fan jets is either 110° or 80°, which dictates the width each jet covers, and hence their spacing on the spray boom. The 110° are normally spaced 500 mm apart, and the 80° 450 mm apart.

Most jets have a fixed orifice, but the hollow cone jets on small hand sprayers and some air blast machines can be varied by altering the distance between the swirl device and the outer disc, and so altering the liquid througput. This also alters the discharge shape and the drop size.

Output of fixed aperture jets can be varied only by changing the liquid pressure, but the effect is minimal and has a much greater effect on the droplet size. The output of a jet is proportional to the square of the pressure; doubling the liquid pressure increases the flow by only $\sqrt{2}$ or 1.4 times.

Liquid throughput is normally determined by the orifice size (table 4.3). The jet 'number' refers to both the spray angle and the jet size, but is specific to each manufacturer. For example, the TeeJet 11004 produces a 110° angle and has an 04 sized hole; the Lurmark F110–50 also produces 110° and has a nominal volume flow rating of 50 under Lurmark's classification.

Table 4.3

Example of jet output figures

Tip No.	*Liquid pressure (kPa)*	*Capacity 1 nozzle (l/min)*	*Litres per hectare 50 cm spacing*				
			6 km/h	*8 km/h*	*10 km/h*	*12 km/h*	*14 km/h*
11003	150	0.84	167	126	100	84	72
	200	0.97	193	145	116	97	83
	250	1.08	220	162	130	108	93
	300	1.18	240	178	142	118	101
	350	1.28	260	192	153	128	110
	400	1.37	270	210	164	137	117
11004	150	1.12	220	167	134	112	96
	200	1.29	260	193	155	129	111
	250	1.44	290	220	173	144	124
	300	1.58	320	240	189	158	135
	350	1.71	340	260	200	171	146
	400	1.82	360	270	220	182	156
11005	150	1.40	280	210	167	140	120
	200	1.61	320	240	193	161	138
	250	1.80	360	270	220	180	154
	300	1.97	390	300	240	197	169
	350	2.13	430	320	260	210	183
	400	2.28	460	340	270	230	195

Source: Spraying Systems Ltd (TeeJet).

Jets are carried in special mountings which allow easy changing without tools. It is becoming increasingly common for the jets to be set into a colour-coded cap which simplifies identification of jet size.

Droplet size varies with pressure, finer droplets being produced by higher pressures. Most jet manufacturers offer 'high-pressure' or 'low-pressure' jets, the latter being capable of producing a high output of large droplets at low pressure. The normal pressure range of jets is

Low-pressure fan 70–275 kPa
High-pressure fan 140–520 kPa
High-pressure cone 280–1100 kPa

Some wettable powders and liquid fertilisers can block small holes, so large-orifice, low-pressure jets have to be used to aid the passage of materials.

4.2.2.2 Basic components of a sprayer
In addition to the jets, the important components are as follows.

(a) Tank. This holds the liquid. Its essential features are non-corrosive material, a large, easily accessible filler lid, and a shape that prevents areas of poor agitation.

Some anti-corrosion treatments that are proof against water corrosion, such as galvanising, can react with certain chemicals and have to be avoided. Tank capacity determines the weight of the sprayer – 1000 litres will weigh around 1 tonne, and a 20 litre knapsack will weigh 20 kg.

(b) Agitator. This is in the form of a mechanical paddle or else agitation is produced by excess liquid returning from the pump. The action produced must be capable of keeping powders in suspension, but must also be controllable to avoid creating excess froth in some liquids.

(c) Pump. This moves liquid, and creates the necessary liquid pressure. Three types are used: roller vane, diaphragm and piston (figure 4.5).

The roller vane pump is susceptible to wear on the rollers and case clearances, which causes performance to fall slowly. The diaphragm does not suffer from wear. If anything fails, the pump stops working. The piston pump produces high pressures, and is lined with hard ceramics to prevent wear and corrosion.

The piston or diaphragm action produces a pulsating flow, which has to be smoothed by the 'damper'. This consists of a hollow sphere, divided by a flexible diaphragm; the lower half is connected to the liquid line and a charge of air is trapped in the upper half. As the air charge is compressible, it is able to absorb the liquid pressure fluctuations. A tyre inflation valve is fitted to the top of the sphere, so that the air charge can be adjusted for the system operating pressure.

(d) Regulator. The pump provides more liquid than is required by the jets, the excess being returned to the tank by means of an adjustable regulator. The commonest type maintains a set liquid pressure flowing to the jets by balancing the jet line pressure against a spring; the pressure on the spring can be altered to vary the liquid pressure.

Other liquid flow systems have been tried. One, the 'Regulo-flo', is a type of flow divider for use with a piston or diaphragm pump, it consists of a disc with a hole which allows the excess liquid flow to return to the tank. Liquid flow through the disc hole is proportional to the flow through the jet holes. Once the correct disc has been inserted, the jet flow rate will remain proportional to the PTO speed, and therefore forward speed, while the same gear is engaged.

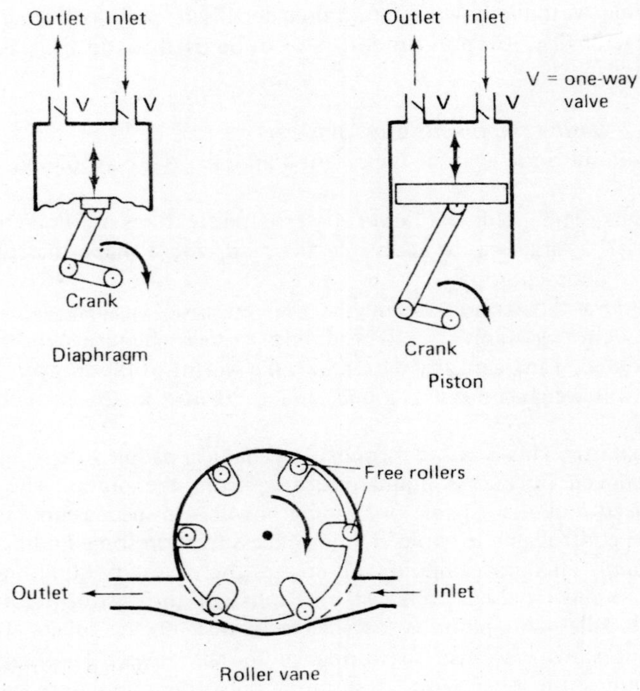

Figure 4.5 Sprayer pumps.

(e) Pressure gauge. This shows the pressure of the liquid being delivered to the jets. Most gauge systems have a means of preventing the chemical reaching their transducer mechanisms. The gauge is also 'damped' to reduce needle fluctuation caused by vibration.

(f) Valves. There are several valves in the sprayer pipework. The main one is that which switches the flow, and is often combined into a multi-parted valve which enables all the pump flow to circulate to the tank for mixing, or for pump filling the tank from an open supply.

Another set of valves controls flow to individual boom sections, often splitting it into three (centre and either side) which allows partial bout widths to be sprayed. With the advent of sealed tractor cabs, these valves can now be obtained with remote electrical operation.

(g) Boom. This supports the jet assemblies, and its function is to hold them at a fixed level above the crop. Booms can be up to 18 m wide, and present serious problems in achieving an even height as a small movement at the sprayer wheel is magnified to a large movement at the boom end. To obviate this, the boom is not fixed to the sprayer chassis, but has a complex arrangement of arms and pivots which allow the chassis to move independently of the boom. The boom cannot be freely mounted, otherwise it would oscillate independently, so a series of dampers is fitted to restrain its movement. The boom is mounted on a vertical slide to allow its height to be varied. It folds for road travel, and also incorporates safety catches to allow it to 'break back' if it hits an obstruction.

(b) Boom pipework. The pipework must be capable of carrying liquid to all the jets without significant pressure drop. This can be helped by feeding the liquid into the centre of pipe runs, rather than into one end.

The jet assemblies are fitted to the boom pipe. Most assemblies contain a filter, to prevent debris from clogging the jets, and can also contain an anti-drip valve, a spring-loaded ball or rubber diaphragm which is forced off its seat by supply liquid pressure, but falls back to seal against gravity flow when the pressure is switched off. This prevents liquid dribbling out and causing damage when the sprayer switches off. It also keeps the boom full to allow instantaneous flow when it is switched on again. An alternative system for anti-drip used to be reversing the flow on the main control valve so that the pump sucked out the boom contents. Thus emptied, there was a delay between switching on and the jets spraying.

Some sprayer booms have two pipe runs, each fitted with jet holders, which enable two sets of jets of different output to be fitted. Two rates can be selected by switching on the appropriate line, and a third rate by using both lines. An alternative to this is a multi-jet assembly which can hold 3 or 4 jets; this can be turned to bring the appropriate jet into operation.

4.2.2.3 Automatic control

Many systems have been designed to automatically link application rate with forward speed. These normally embody a speed detector, a rate computer and a flow controller. Speed is measured by either radar or counting wheel revolutions. A programmable computer unit combines this with the jet flow characteristics, to select the pressure at which the required output is achieved. The controller regulates either flow or line pressure. The system is able to regulate flow to speed, so that overall application rate remains constant. The main disadvantage is that altering the flow or pressure through a jet will alter the droplet performance; at slow speed, large drops will be produced, and high speed will produce fine mist. Some computers recognise this problem and will adjust output within limits; outside these the driver is warned that his speed is wrong. Most computers will also show field area application rate forward speed.

4.2.2.4 Concentrate injection

The sprayer operates on water only, with the chemical being metered into the flow just before the jets. The chemical is handled by its own positive displacement dosing pump, driven directly from a ground wheel. This benefits both accuracy and safety because

(a) The chemical does not have to be mixed into the tank water and agitated.
(b) The chemical can be supplied in a special container which plugs directly into the sprayer, so reducing the risk of handling accidents.
(c) The sprayer can apply a constant rate of water to suit the jet performance, with concentrate being metered in to suit forward speed.

4.2.2.5 Electrostatic deposition

The problems of insufficient gravitational force on small droplets can be overcome using electrical attraction. If a spray droplet is electrically charged, it will be attracted to any object at zero potential (that is, an 'earthed' plant). The droplets are charged by creating a high-voltage field around each jet. The charges involved are high – 4000-30 000 volts – but the current required is too low to render a spray charging head dangerous to the operator. The main problems are in providing sufficient insulation to prevent the voltage from leaking to earth via the sprayer chassis.

4.2.3 Rotary atomisers

These machines use a high-speed (up to 8000 rpm) spinning disc to break the liquid droplets. The liquid is fed on to the centre and moves across the disc face in a thin sheet. The sheet breaks into drops at the disc edge; sometimes teeth are added to aid break-up. Another atomiser uses a rotary cage where the drops are formed as the liquid is flung from peripheral bars. This system can produce almost uniform droplets, the size being determined by the disc speed and the liquid flow, and can use concentrated, specially formulated chemicals. Rates as low as 10–15 litres/ha are possible, instead of 300 litres/ha or over on hydraulic systems.

The atomiser disc drive requires very little power, and a single head version, which runs on torch batteries, is available for small areas. Larger output single head units are used on top of tractor-mounted masts to drift spray field chemicals.

These systems are used in multiple head versions mounted along a boom. They normally incorporate slower discs or cages to apply outputs nearer those of conventional hydraulic machines, but still within a close droplet size range. The discs are driven hydraulically or mechanically, liquid being fed by a small pump and pipe distribution system to each disc. Discs are mounted horizontally or vertically. On both types only part of the full discharge circle is used, by enclosing the sector where discharge is not required, and catching the liquid for return to the tank.

4.2.4 Specialist horticultural sprayers

Many crops have special chemical application requirements which are met by purpose-built machinery.

4.2.4.1 Band sprayers

Band spraying is used to apply a chemical only over the plant row. Band spraying on a seeder is by a single nozzle behind each unit in established crops. One or more nozzles are mounted over each row on a simple toolbar. Normally cone jets are used to achieve an even cover across the width of the sprayed band. The band sprayer uses the same basic components as a normal sprayer except for the nozzle mounting arrangements.

4.2.4.2 Strawberries

Dense foliage rowcrop sprayers are used for strawberries. In the simplest form it consists of an arched framework, over the plant row, holding several inward pointing jets to ensure that spray is applied from both sides as well as from above. Better application is achieved by ruffling the leaves as the liquid is applied, by a length of light chain loosely hung across the arch.

4.2.4.3 Droplegs

For sprouts or raspberry canes the spray has to be applied to the sides, rather than the top, of the crop. Therefore the jets are attached to vertical 'dropper legs' hanging below the transverse boom. The brussels sprout machine has specially shaped droppers which gently part the leaves and lift them to allow the spray to reach the buttons. Some conventional booms can be converted to drop let.

A single row bush fruit version has been made, consisting of a hand-pushed trolley with tank and pump, and with vertical booms either side.

4.2.4.4 Inter-row herbicides

Total herbicides can be used between row crops in place of hoeing. A band sprayer is used for this, but the crop has to be protected from droplets by metal shields, which are normally free floating to allow them to follow the ground closely. Drift problems are reduced by using low-pressure, large-droplet jets.

When spraying between strawberries, discs must be used ahead of the shields to sever runners and prevent them from translocating spray back to the mother plants.

4.2.4.5 Small areas and glasshouses

Larger-capacity machines for market gardens and sports fields are hand or light tractor pulled versions of the field sprayers discussed above. Boom widths are up to 3 m. The tank and fittings are mounted on a light chassis with a small petrol engine to power the pump.

Some sprayers used in glasshouses consist of a tank and electric pump on a trolley, the hand lance being fed via a long hose.

Most small horticultural sprayers of up to 10 litres (2 gal) capacity use air pressure to expel the chemical from the tank, after being charged with a hand pump or compressed gas. The simplest ones feed directly to a hand lance, without control of pressure or output. Better models have a regulator in the discharge pipe to control the pressure to the jets. Some knapsack sprayers incorporate a hand-operated liquid pump. The pump incorporates an air chamber which stores pressure to maintain spraying on the pump return stroke. The chamber also allows flow regulation without complex plumbing.

Portable 'mistblowers' are versions of a knapsack sprayer which use a small petrol engined fan to blow the atomised spray into the crop. They are backpack-mounted and consist of a liquid tank, engine-driven fan and spray metering/atomising system. The spray is discharged through a hand-held nozzle. Some versions can also apply dry powders.

Where portable sprayers are used in row crops or beneath bushes, specially shielded lance heads are fitted to avoid harming the plants. It is also possible to fit a simple wheeled shield assembly for long crop rows.

It is often difficult to maintain a fixed 'travel speed' when walking with a pedestrian-operated sprayer. A small electronic metronome has been devised which can be set to bleep in time with the pace needed to provide the correct rate.

4.2.5 Electrodyne

This atomises the chemical stream by electrically charging its particles. The charged particles repel one another so violently that they are thrown out into a vapour cloud. Although high charge voltage is needed, power consumption is very low, and a unit can be hand held and operated on batteries. The chemical must be suitable for electrically charging, and this dictates a water base rather than the more common oil base.

4.2.6 Air-assisted sprayers

These are used to apply liquid chemicals to dense foliage, such as bush and top fruit orchards. A mist of droplets is generated by hydraulic nozzles and blown into the foliage.

The airflow and droplet size must be sufficient to project the chemical into the crop canopy. Most chemicals must coat both sides of the leaves to be fully effective.

Research by the N.I.A.E. has shown that airspeed can affect the coverage of individual leaves. Their findings for blackcurrant bushes are shown in table 4.4.

There will be similar effects, though at different airspeeds on most other orchard crops. The critical settings for any machine are found with tracers and leaf analysis.

Table 4.4

Airstream effects on blackcurrants

Airspeed (m/s)	*Effect on leaf*
Under 12	Leaf remains still, droplets hit one face only
12-15	Leaf gently flutters, exposing both sides to the spray
Over 15	Leaf acts like a weather vane. Lies along airstream presenting narrow edge only

4.2.6.1 Atomisers

Liquid atomising is very similar to a hydraulic machine. The pump is diaphragm or piston, the latter being more common in order to achieve the high nozzle pressures of 400–700 kPa. As the chemicals are often powder suspensions, heavy agitation is needed.

The jet types are hollow cone, either replaceable orifice or adjustable, the latter enabling a nozzle to be shut down by means of a screw. The nozzle banks are placed in the fan discharge so that a combination of high velocity and turbulence thoroughly mixes the droplets into the airstream.

4.2.6.2 Fans

The airblast is created by powerful fans. On most machines the discharge is at right angles to travel (figure 4.6), although some have an adjustable 'cannon' air nozzle to blow a concentrated stream into tall foliage or across the tops of vineyards.

Three types of fan are used to create the airblast, each with its own attributes:

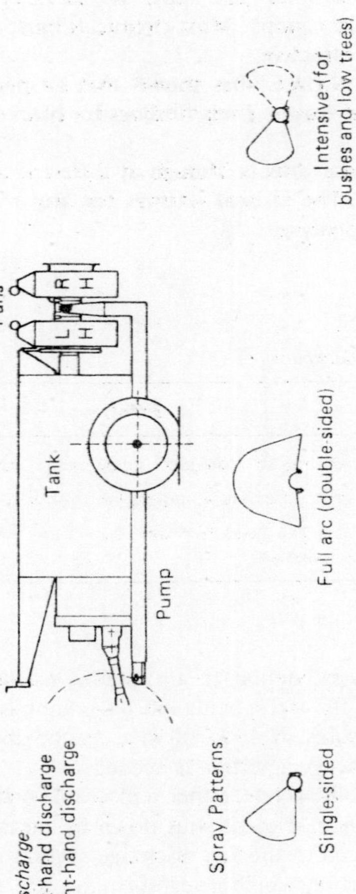

Figure 4.6 Orchard sprayer.

(a) Axial flow. These are 'propeller' type fans with their axes along the line of travel. Immediately behind the fan is a deflector cone to turn the airstream outwards. The airstream passes through further straighteners and deflectors to give the desired spread pattern. On most machines the final deflectors can be adjusted to suit the required spread pattern and orchard layout. This arrangement is mechanically simple, but absorbs power in deflecting the fan discharge. A typical unit will absorb 10–20 kW.

(b) Centrifugal. These have their impeller shafts along the line of travel, and discharge sideways. Deflector plates in the discharge produce the required tall, thin airstream, but there are limitations to the degree of deflection possible, so one fan will blow only to one side of the sprayer. For spread to both sides, two fans are needed, one discharging at each side. The air deflector plates are easily adjusted to give the required spread pattern. Deflection is not as severe as in the axial fan, so less power is used.

(c) Tangential flow. These are similar to centrifugal units but have very wide (1 m), small diameter (300 mm) impellers. When mounted with its shaft vertical, the fan produces an airstream of the correct shape without the need for deflectors. As there is no power consumed in turning or deflecting the airflow, drive requirements are under 50 per cent of the axial unit. One fan is needed per side for bush fruit, and a further number at varying angles would be needed for full arc spraying of top fruit.

To satisfy the need for varying airspeeds into the crop, fan outputs have to be adjustable. This is normally done by varying the speed, either at the PTO or, where fitted, by variable hydraulic drive. The direct PTO machines cannot be slowed below the point where the pump is unable to supply sufficient spray. To reduce the airflow further, fan inlet restrictors have to be used, for instance on sprayers with adjustable spread patterns, when set to discharge in a limited arc.

4.2.6.3 Orchard sprayers

The confined pathways in orchards require good manoeuvrability. Trailed machines have to be capable of sharp turns, and so require special PTO drives which can run at acute angles. Some sprayers have a split construction, with the fan and pump unit mounted on the tractor linkage, and the tank being towed behind.

4.2.7 Fogging

This involves generating a vapour from a liquid or, in some cases, from crystalline solids, which can pass into the atmosphere around the crop. Four systems are available:

(a) Heated pots, which consists of an open topped cup with a small electric heater in the base. Special liquid or crystal formulations, which are slowly vaporised by the heat, are used inside buildings, and the vapour spreads into the atmosphere by diffusion.
(b) Mechanical atomisers are similar to air-assisted sprayers except that the droplets pass through fan blades which shatter them into fine particles, and blow them into the atmosphere.
(c) Engine vaporising foggers use a small petrol engine to drive a turbine in the liquid tank. The turbine heats the liquid by friction, and also pumps the hot liquid into the engine exhaust pipe, where it vaporises. The vapour and exhaust gases are blown into the atmosphere by pressure from the engine, without additional fans. The vapour is produced at 120–150°C by this process.
(d) Pulse jet foggers inject the chemical into a fuel/air mixture in a chamber. Upon ignition the chemical is vaporised, and the explosive forces push the vapour out of the chamber. The chamber shape is designed so that, as the pressure drops after the explosion, a fresh charge of fuel, air and chemical is induced, without needing a piston or crankshaft. Ignition is achieved by an electric spark plug, triggered by the induction process. The machine runs at around 80 pulses per second. The chemical is heated to over 500°C, and some water-based carriers cannot be used. Its main advantage over the engine type is that there are no moving parts and the weight is less.

4.2.8 Contact applicators

These have been designed to apply total herbicides to weeds projecting above crop level. On grassland this treatment can kill docks, thistles and nettles; in root vegetable crops, plants that have run to seed are also killed. The essential aspects are to apply sufficient herbicide without it dripping on the crop beneath. Three systems to do this have been devised.

(a) Herbicide moistened 'wicks' consist of a hollow metal boom up to 12 m wide, which contains concentrated chemical. The wick is made of small lengths of special rope attached along the boom, with both ends pushed through the boom wall into the liquid. On some machines, liquid flows to the rope by gravity and capillarity, while others slightly pressurise the boom. The rope material allows liquid to leave it only by contact and not by gravity, and thus flow stops when wiping stops. To ensure cut-off, some booms can be rolled to position the rope outlets upwards.

To ensure sufficient transfer at a reasonable speed, various configurations of wick have been tried. Some use double runs of straight wick, while others push the centre of each rope forwards to form a series of Vees through which the weeds are drawn.

With few exceptions the machines require no power, and are light and easily handled. A small unit 600 mm wide can be hand held, and larger units can fit across the front of any vehicle.

(b) The carpet wiper has a moistened strip of matting which forms the front face of the boom. Chemical is sprayed on to the inner face of this matting by a small pump and hydraulic nozzles. Excess chemical drains down the inner face of the matting, and is collected in the base of the boom for return to the tank. This unit offers a greater contact area and positive wetting.

(c) Wipers with mechanical action have been made with a pair of full width rollers. One runs in a shallow tray containing the chemical, which it picks up and transfers to the other roller in an even coating. The weeds rub the second roller. As in previous types, the chemical layer on the second roller must not be so thick as to drip off, but it must nevertheless readily transfer to the weeds.

4.2.9 Sprayer operation and calibration

Sprayers must be checked regularly to ensure efficient application. This entails two separate operations: first to see if it is operating correctly, and second to check the application rate.

Although all the following checks are carried out with water, there is still a risk of it becoming contaminated with chemical and full protective clothing must be worn.

4.2.9.1 Operation

(a) Jets. It is essential that each jet delivers the correct amount of liquid, and in the required pattern. The output variation between a series of jets can be checked by allowing each one to spray into a small glass container for a set time, normally between 20 and 60 seconds. Ideally this should be a measuring cylinder but, if not, the liquid depth in the container for each jet can be measured with a rule. The jets should not vary by more than ± 5 per cent of their average output. For example, if the average output in 30 seconds was 500 ml or 100 mm depth, then those giving under 475 ml or 95 mm, and those giving over 525 ml or 105 mm, should be renewed. A more rapid check can be obtained with a proprietary flowmeter designed to fit beneath each jet in turn. This reads directly in litres/second.

If the output progressively falls along a section of the boom, it will indicate excessive pressure drop owing to inadequately sized plumbing.

Each jet must be checked visually for correct operation, and any not producing a uniform droplet pattern must be renewed. Fan jets must be aligned so that they are just out of parallel with the boom, so as not to impinge on each other.

Blocked jets must not be cleared by mouth or poking with metal; compressed air can be used.

Poor jet performance can be caused by blocked jet filters or sticking anti-drip devices.

(b) Operating height. Fan jets must be set to a height where the triangular discharge from one meets that from the next but one jet at the crop surface. Cone jets and other band-spray nozzles should be at the height to provide the band width. This can be checked by spraying on to concrete to see the width of the wetted area.

On air-assisted sprayers, the discharge pattern must match the height and shape of the crop being sprayed.

(c) Pressure gauge. This is an important part of operator setting. Any gauge that does not return to zero or that fluctuates widely must be changed. If the output calibration does not appear to conform to jet flow rates at the stated pressure, the gauge is suspect.

4.2.9.2. Calibration

Checking application rate is a simple task, involving measurement of the amount of liquid used to spray a known area. For field scale machines this is best done over a distance equivalent to 0.1 ha; the travel distance that represents 0.1 ha can be calculated as follows

$$\text{travel distance (m)} = \frac{1000}{\text{boom width (m)}}$$

For small horticultural and knapsack sprayers the test area can be based on 100 m^2.

The liquid consumption is measured easily by completely filling the tank, spraying the test area only, and measuring the quantity of water needed to refill it.

The sprayer should be run across the test area at the intended speed and gear. The tractor should be up to speed before it reaches the start, and not stop until it is beyond the end of the measured plot.

Example 1

A 12 m machine is required to apply 330 litres/ha.
Travel distance for 0.1 ha = 1000/12 = 83 m.
Thus the sprayer must discharge

$$330 \times 0.1 = 33 \text{ litres in } 83 \text{ m}$$

It is sometimes easier to calibrate small sprayers in terms of millilitres (ml) per square metre; to convert to this rate from l/ha, the factor is

$$\frac{\text{litres/ha}}{10} = \text{ml/m}^2$$

Example 2

A knapsack sprayer, applying the equivalent of 330 l/ha will deliver

$$\frac{330}{10} = 33 \text{ ml/m}^2$$

4.2.9.3 Concentration dilution rates

The amount of concentrate to be added to the tank is calculated as follows.

Concentrate litres/100 litres of tank capacity =

$$\frac{100 \times \text{concentrate rate/ha}}{\text{overall application rate/ha}}$$

For horticultural sprayers, concentrate ml/litre of tank =

$$\frac{1000 \times \text{concentrate rate/ha}}{\text{overall application rate/ha}}$$

Example 3

The sprayer in example 1 has a 700 litre tank, and needs to apply 6 l/ha of concentrate. Therefore each full tank will need

$$\frac{100 \times 6 \times 7}{330} = 12.7 \text{ litres of concentrate}$$

Example 4

The knapsack in example 2 holds 20 litres and is to apply the same rate as in example 3. Therefore each full tank will need

$$\frac{1000 \times 6 \times 20}{330} = 364 \text{ ml of concentrate}$$

4.2.9.4 Changing rate

This is accomplished by

(a) Changing pressure – small variations only; this can alter the droplet spectrum if changes are too severe.
(b) Changing jets – large variations; this might need to be combined with pressure change in order to retain the desired droplet spectrum.
(c) Changing the forward speed – normally only large variations owing to the gear ratio steps.

4.2.9.5 Checking coverage

Two visual methods can be used to ascertain the uniformity of application on crop surfaces.

(a) Fluorescent tracer. A fluorescent dye, normally yellow or red, is added to the spray. After spraying, samples of plants are collected and viewed under ultra-violet light, where spray deposits will show up as illuminated spots.

Ultra-violet light is harmful to the eyes, so the lamp must shine on the samples only, and be shielded from the direct view of the tester.

(b) Water-sensitive paper. This is specially treated paper, normally yellow when dry, which turns blue when wetted. The paper treatment also prevents lateral water travel, so that a droplet hitting it will only colour an area corresponding to its volume. After spraying, the paper is covered with blue dots of different sizes which can be visually assessed for uniformity, and sizes of droplets deduced.

In use, small pieces of the paper are clipped to the plants to be sprayed. The main disadvantage in this method is that the weight of the paper and clip can change the attitude of a leaf, especially when used for an air-assisted machine.

4.3 SOLID PESTICIDES

These are in the form of granules or dusts, made by chemically impregnating an inert carrier material, for instance limestone, sand, clay or even coal. They are chosen for their chemical compatibility, rather than their compatibility with moving parts of machinery, so problems can arise from factors like abrasiveness.

4.3.1 Dusting machinery

(a) One group of machines operates by blowing a fine suspension of dust among the crop. The discharge distance varies between 5 and 50 m, depending on fan power and dust particle size. The powder is metered into the airstream by either of two methods.

In the first method, the powder is contained in an airtight hopper. A powerful air jet in the base, blowing through the powder, creates a cloud of dust in the hopper top which is drawn off by the main airstream. The metering rate is controlled by the power of the agitation air jet, but is very dependent on the powder material.

The second method uses a variable-speed auger or metering wheel to carry powder from the hopper to the airstream.

Dusters can be obtained in a range of sizes. Small, hand-held or pedestrian barrow types with petrol or mains electricity motors can work inside buildings. Large tractor-mounted versions are made for field-scale operations.

(b) Crop products like potatoes can be dusted as they travel along a conveyor. This is done by dropping the powder from a hopper above the conveyor. It is metered by a variable auger or vibrating chute, at a rate commensurate with the crop flow on the conveyor.

4.3.2 Granule applicators

Most granule dispensers meter by an external force-feed rotor in the base of a hopper. The metering rotor has flutes to suit the product granule size (figure 4.7). Discharge rate is controlled by the rotor drive speed, and adjusting the width of the rotor face used for metering. The rotor requires little power, and several can be driven by a light, spoked wheel running on the ground. As this wheel does not have to support any equipment, it can be lifted off the ground in order to stop the flow. The metering rotor material and flute shape have to be chosen so as not to damage soft granules but to resist wear from abrasive materials. The hoppers hold only 5–10 kg of material, but this is commensurate with the low application rates normally used.

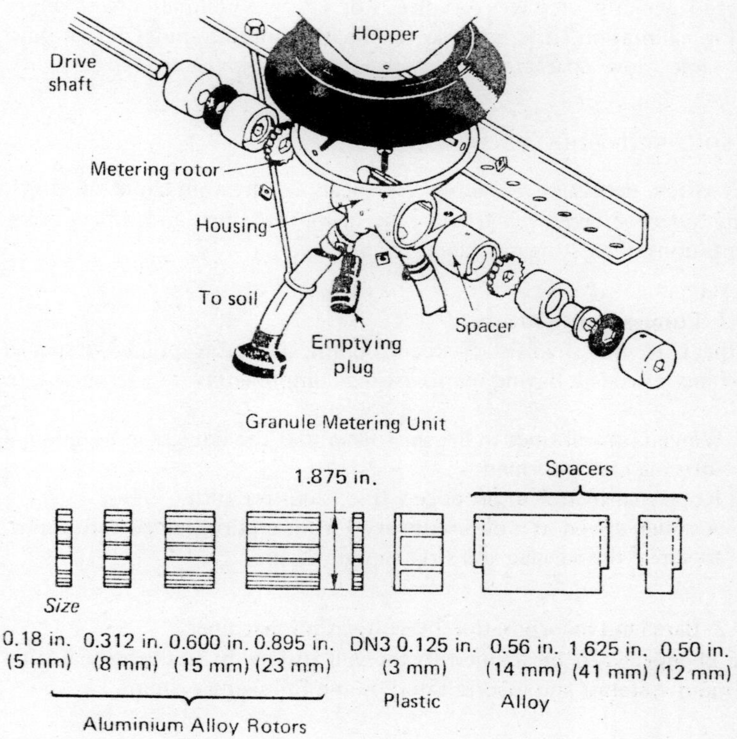

Figure 4.7 Granule dispener.

Many machines allow the granules to run down tubes to the soil. The bottom of the tube can have a variety of discharge fittings to produce the band size required, or even to incorporate the granules to a shallow depth into the soil. For overall application, fan-assisted applicators are used. As in the pneumatic fertiliser spreader, the granules are blown against deflector plates to achieve the scatter.

The simple hopper units are easily mounted on drills and planters, or on their own frame. One frame layout is designed to be sandwiched between the tractor linkage and the drill or planter headstock.

Granular material like lawn sand can be spread with many of the fertiliser distributors described at the beginning of this chapter.

Slug pellets, which require very low rates, can be spread with fertiliser equipment if its discharge rate can be reduced sufficiently. The small, electrically driven spinning disc described in chapter 3 is also suited for slug pellet dispersal.

Calibration of granule applicators is similar to that for seed drills, and the same methods apply to calculating the weight of material discharged per unit of travel distance. For safety when handling granules during calibration, it is possible to purchase inert granules which have the same flow characteristics without containing active ingredients.

4.4 SOIL INCORPORATION OF PESTICIDES

It is often necessary to apply chemicals to the root zone of crops. Many systems have been tried, some being specialist, and others being adaptations of existing machinery.

4.4.1 Fumigation sterilants

In the U.K. this involves metham-sodium, a volatile liquid. A special machine is needed, having the following components

(a) Winged subsoil tines to lift the soil so that the liquid can be sprayed into the cavity formed.
(b) Rotary cultivator to produce a fine, clod-free tilth.
(c) A roller driven at a different speed from that of the forward travel, to smear the surface and stop vapour escape.

4.4.2 Bare land incorporation by cultivation machinery

The chemical can be applied to the soil surface by conventional solid or liquid systems, and incorporated by the following systems.

(a) Tine harrows. These are unable to get material lower than 50 mm and are poor at lateral mixing, so they need the material to be evenly applied to the surface.

(b) Vertical axis, rotary cultivators. These are suitable for incorporating to 75–100 mm, with potentially good lateral mixing. The applicator can be put on to the cultivator, but the point of application is critical, otherwise the chemical can be left in bands corresponding with the points where the rotors run together.

(c) Horizontal rotor cultivators. These can incorporate to 150 mm with poor lateral mixing, but this is overcome by applying a full width band into the front of the rotor.

(d) Special deep incorporators. These are special cultivators with subsoil type tines. The applicator discharges down tubes behind the vertical leg.

4.4.3 Drill or planter incorporation

Many soil-acting pesticides are incorporated with the plant or seed for maximum effect. One of the main problems with this technique is that the chemical can end up in the wrong position with respect to the plant roots. This manifests itself as either non-effectiveness, because the chemical is too remote from the plant, or toxicity, because the chemical is too close to the roots. Chemicals are normally in granular form, and are metered out of the hopper units described in section 4.3.2. The methods of soil application are

(a) A surface band ahead of the drill/planter. This assumes that the chemical will be incorporated by the soil disturbance caused by the coulter. In practice, some coulters can move the band aside an opening, and the coverers fail to recover any part of it. Where a narrow band is laid by an applicator on the front of the planting tractor, there is a danger of the coulter missing the band completely owing to steering inaccuracies.

(b) Root zone-acting chemicals run into the furrow with the plant or seed. If the chemical is toxic to the plant by direct contact, the applicator tube can be moved slightly behind the seed or plant drop point. This will allow the chemical to mix into the soil as it is returned by the coverer.

(c) Stem protection chemicals. These need to be worked into the soil around the stem. The system described in (b) above can be used, but the 'bow wave' method is more successful. This is a separate applicator coulter, immediately ahead of the main coulter, which allows the granules to mix into the band of soil in which the main coulter works, so ensuring that all the soil moved back around the plant will contain chemical.

(d) Spot application. In widely spaced crops much of a band is wasted. Only the area around the plant is of interest from the application viewpoint. While this is theoretically feasible, it is difficult to synchronise to the seed or plant drop on a planter, although some positive placement planters have trigger facilities, which operate in phase with plant spacing. Post-emergence spotting is again difficult to synchronise. Even plant position sensing is difficult. It is possible, however, to get a hand machine where a small dose of granules can be dropped at each plant. Operation is either by a small battery-powered solenoid and press switch, or by jerking or tilting the dispenser which causes a piston to meter a dose.

5 IRRIGATION

5.1 SYSTEM DESIGN

5.1.1 Crop requirements

A prerequisite to selecting the equipment for an irrigation system is calculation of the amount of water to be applied. Three factors are required in this calculation: precipitation, frequency and crop area.

(i) Precipitation is measured in terms of mm of water, equivalent to the way in which rainfall is quoted. The figure for any crop is derived from complex calculations involving plant leaf area, likely crop response, and meteorological data, and these vary considerably from season to season. Sufficient data have been collated for most crops to enable guidance rates to be formulated. However, to design a system, only the likely maximum precipitation rate will be required.
(ii) Frequency interacts with precipitation rate to determine the amount of equipment needed, its layout, and the water supply requirements. In multiple crop situations, the frequency and application timings for each crop must be combined to see if the system has to irrigate more than one crop area at once.

For example, a crop of outdoor lettuce might want an application of 25 mm of water every five days in mid-summer, whereas a raspberry crop might need 50 mm at one stage of growth only. Thus, given the same area of each, a lettuce system could require less equipment, but a larger reservoir.
(iii) Crop area will determine the total quantity of water needed and, when precipitation and frequency are taken into account, the amount of equipment needed, and water supply rate.

Volume of water is often quoted in terms of 'hectare millimetres' ('acre inches'), and is a measure of the quantity of water needed to apply a rainfall equivalent of 1 mm on a hectare

1 ha mm = 10 000 litres, or 10 m^3

For example, 1.5 ha of lettuce will require 375 000 litres of water to apply the equivalent of 25 mm of rain.

The total amount of water required per year or per season is calculated by totalling the seasonal irrigation requirements and area of each crop. This figure is used for planning a holding reservoir or obtaining an abstraction licence (table 5.1).

Table 5.1

Total water requirements of a holding

Crop	*Area (ha)*	*Seasonal precipitation requirement (mm)*	*Seasonal water requirement (litres)*
Lettuce	4	125	5 000 000
Raspberries	0.5	50	250 000
Potatoes	10	100	10 000 000
Cauliflower	8	35	2 800 000
		Total =	18 050 000

5.1.2 Water flow rates

When the volume of water is divided by the time that the irrigator will run, the flow rate of the water supply is found. For example, if the lettuce crop in the previous example is irrigated in one block, over 10 hours, the supply flow rate will be

$$\frac{375\,000}{10} = 37\,500 \text{ l/h or } 10.4 \text{ l/s}$$

5.1.3 Storage requirements

5.1.3.1 Quantity

Often, all or part of the water used for irrigation has to be held in a tank or reservoir. Two systems are used, full storage or back-up storage. Full storage is used where there is no available water source during the irrigation season, so that the total volume must be held over from winter rainfall. Back-up storage is needed where the water supply flow rate is less than the application rate, but can catch up during the periods when the system is not working.

Full storage requirements are calculated as for total volume as in section 5.1.2.

Back-up reservoir capacity is calculated by taking the volume of water to be applied during one setting, and subtracting from it the amount of water that can be supplied during the same period. For example, a series of 1.5 ha blocks of lettuce are to be irrigated, one per day, but the mains will supply only 20 000 l/h. For total supply the reservoir would need to hold 375 000 litres, but during the 10 hour application period, the mains will be replacing 20 000 × 10 = 200 000 litres. Thus the reservoir will need to store only 375 000 – 200 000 = 175 000 litres; this amount of water will be replaced during the 14 hours when the irrigation equipment is not working.

Even where the mains supply is adequate, many water authorities stipulate the use of a 'break tank', where the mains water runs into a tank, from which the pump draws. Attaching a system directly to the mains has the risk of allowing nutrient additives to backfeed into the main or the pump, drawing too heavily on the main and affecting other supplies.

5.1.3.2 Reservoir design

(a) For tanks, the capacity is easily decided by calculating the volume as follows.

Square or rectangular

$$\text{volume (litres)} = \text{width (m)} \times \text{length (m)} \times \text{height (m)} \times 1000$$

Cylindrical

$$\text{Volume (litres)} = \frac{\text{diameter}^2}{2} \times \pi \times \text{height} \times 1000$$

(b) Excavated reservoirs appear more complex, as the sides are angled rather than vertical. The area of the base is considerably less than the area of the top, so the first calculation is to determine the average of those areas. In most cases, the average area occurs at half the depth, that is, the water surface area when half full.

$$\text{Average area (m}^2) = \frac{\text{volume (litres)}}{1000 \times \text{depth (m)}}$$

Taking the example from table 5.1, for a depth of 3 m

$$\text{Average area} = \frac{18\,050\,000}{1000 \times 3} = 6017 \text{ m}^2$$

The overall area occupied by a reservoir will depend on the shape of the excavated sides and the banking. To stop the sides collapsing, the excavation is gently sloped (battered). The batter angle depends on the soil type, sand requiring a gentler slope than clay loam. The extent and effect of batter is shown in table 5.2.

For embankments not exceeding 2.7 m in height, the top width should be 2.7 m. The freeboard should be a minimum of 0.5 m.

It will be seen that the battered bank and its top width increase considerably the area required by the water (figure 5.1). Thus the

Table 5.2

Maximum embankment slopes

Soil type		*Reservoir lined with*	
		Butyl rubber	*PVC or Polythene*
Loams Silty loams (lining may not be required)	outside slope inside lined slope	1 in 2 1 in 2.5	1 in 2 1 in 3
Sandy loams Sand Sandy gravels Gravels	outside slope inside lined slope	1 in 2 1 in 2	1 in 2 1 in 3

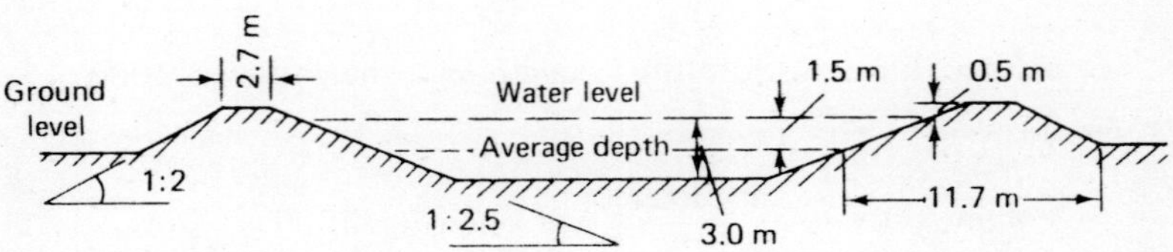

Figure 5.1 Cross-section of reservoir banks.

foregoing reservoir will have an average water surface area of 6017 m^2, or be 78 m square, but the overall land area will be 78 m plus (2 × 11.7)m banks square, or 10 280 m^2 (just over 1 ha).

5.1.4 Distribution pipe size

There is an optimum size for pipework to carry the required quantity of water, to prevent it being grossly over-sized and thus expensive. This is determined by calculation of 'head loss', which is the friction between the water and the pipe sides.

Head loss is measured in Pascals (Pa), although other non-preferred units, such as metres of water or bars, might be quoted.

The calculations for frictional head loss are extremely complex but, for convenience, the data are presented in graphical or tabular forms. The graph in figure 5.2 allows one to convert from water flow rate to head loss for a range of pipe sizes.

The head loss table is specific to one type of pipe; the one shown applies to UPVC (plastic) and aluminium materials, but iron, concrete or asbestos have different internal roughnesses and their resistance would be higher. Note also that the pipe size can be quoted in two ways

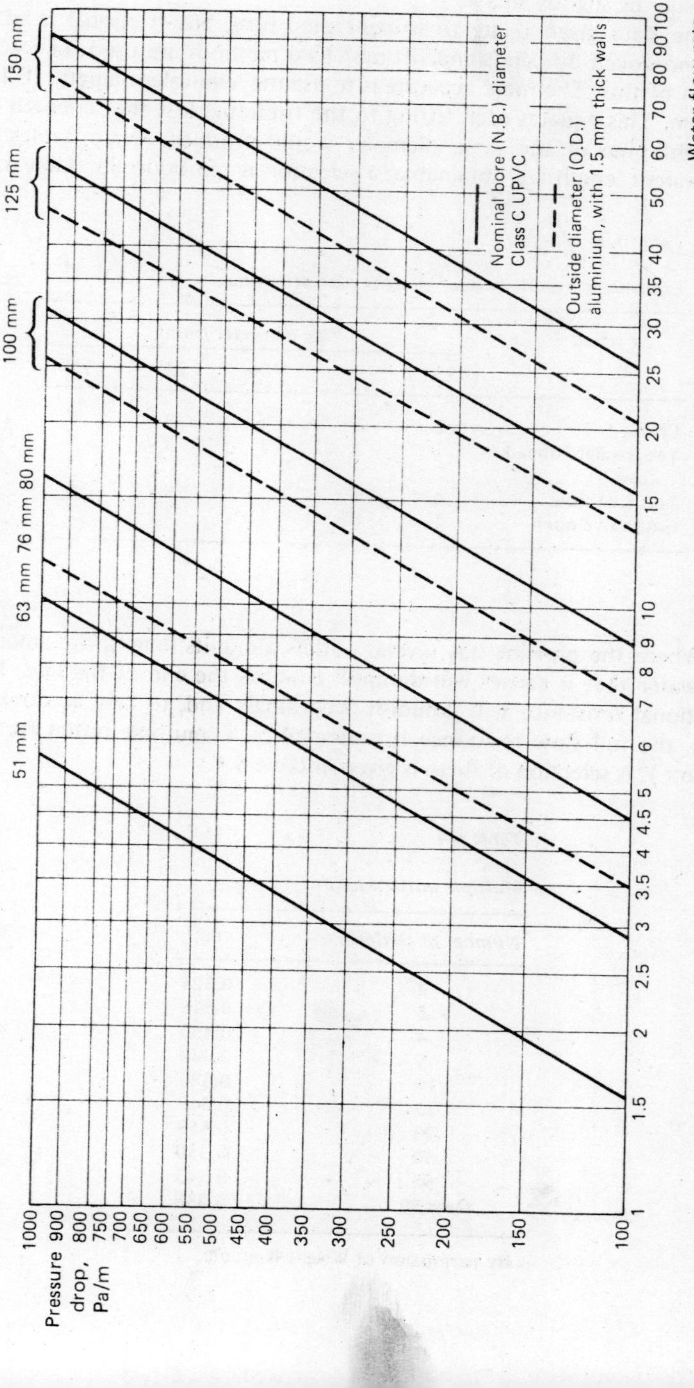

Figure 5.2 Pipe friction chart.

– 'nominal bore' (NB) or 'outside diameter' (OD); the latter pipe will normally be slightly smaller.

The data given relate to *straight* pipe runs, but frictional head is also produced by bends and fittings. Two methods are used for calculation of this. The most accurate is to use the 'equivalent length' (E.L.) system. This equates each fitting to the frictional loss that a length of straight pipe of the same diameter would produce. Again, tables of equivalent length are obtainable; a selection is reproduced in table 5.3.

Table 5.3

Equivalent lengths in m of straight pipe of fittings

Fitting	*Pipe diameter (mm)*				
	50	*75*	*100*	*125*	*150*
Elbow	2	2	3	3	4
Tee, straight-ahead flow	1	2	2	3	4
Tee, side flow	4	5	7	9	10
Gate valve open	1	1	1	1	2

Where the pipeline has several outlets along its length, the amount of water that it carries will diminish towards the end of the line. The frictional resistance will diminish accordingly and, to take account of this, the full flow resistance is corrected by a 'multiple outlet factor' (m.o.f.). A selection of these is given in table 5.4.

Table 5.4

Multiple outlet factors

Number of outlets	*m.o.f.*
2	0.625
3	0.518
4	0.469
5	0.440
10	0.385
15	0.367
20	0.359
30	0.350
50	0.343
Over 50	0.338

By permission of Wright Rain plc.

The alternative method is to calculate the frictional losses for all the straight runs and add 30 per cent to take account of the fittings. This method has obvious advantages when calculating for portable mains, where the exact layout and number of fittings is not known. It often gives a result erring on the safe side, and should be avoided when the exact layout is known as it can result in costly over-sizing of the pump or pipework.

The method for sizing pipework for a scheme is as follows.

(a) Water flow rate: use the maximum likely flow. In systems where the number of sprinklers may vary, calculate on the maximum number.
(b) Obtain data relating to the pipe materials.
(c) Length of run: determine the layout that will require the longest run of pipe and use this figure.
(d) For most systems, a range of mains and portable pipe sizes will be used, and where multiple sprinkler lines are used, the water flow along each part of the system might be different. The total head loss is obtained by calculating the individual parts; this is most conveniently done in a tabular form, like that shown in table 5.5.

Table 5.5

Head loss calculation layout

Run section	*Water flow*	*Pipe size*	*Loss per unit length*	*Section length*	*Fitting E.L.*	*Section head loss*
			Total head loss =			

Where the total head loss is unacceptable, the right-hand 'section head loss' column will indicate those sections that are critical, and will benefit from recalculation with a larger pipe size.
(e) An acceptable figure for head loss will depend on the pump or supply pressure, and the system pressure requirements.

> maximum acceptable frictional head loss = supply or pump head *minus* system pressure needs

The system pressure requirements are normally the irrigator operating pressure and the vertical lift between the water source and the irrigator; these are calculated thus

$$\text{system pressure (Pa)} = \text{irrigator pressure (Pa)} + \text{lift (m)} \times 9.7$$

Available pump head can be found from manufacturers' data for units giving the required flow rate. The available mains pressure figure should be obtained for the supply when it is running.

5.2 WATER STORAGE AND DISTRIBUTION

5.2.1 Storage systems

The storage system is often governed by the required capacity.

5.2.1.1. Tanks

These can be of galvanised steel plate or glass reinforced plastic (fibre-glass). The maximum common size is 5000 litres, small ones of 250–500 litres being suitable as 'break tanks'.

Many tanks of this size can be totally enclosed, except for only an access manhole and pipe connections. This ensures the exclusion of contaminants when clean water is needed on crops for direct consumption.

Prefabricated reservoir tanks consist of sheet metal walls to provide structural support, and an impervious lining. The support walling is commonly a ring of corrugated galvanised steel or welded steel mesh, with butyl rubber sheet for the lining. The tank can be fitted with a tented lid of butyl rubber sheet to keep out leaves and other foreign matter.

Capacity ranges from 3.6 m diameter x 0.9 m high, holding 9000 litres, to 11 m diameter x 2.3 m high, holding 230 000 litres.

Connecting pipework is normally carried over the side, rather than through it, to avoid the need to make watertight joints with the liner.

5.2.1.2 Reservoirs

These are in the form of a large, shallow depression, the excavated soil being formed into banks around the perimeter to increase storage depth. In all but totally impervious clay soils, an impervious lining is needed to make the excavation watertight. Two lining methods are used.

(a) Clay slip. Dry powdered clay is spread in a layer over the inner surfaces. When wetted by the water, it 'puddles' and seals the natural soil pores and small fissures.
(b) Butyl or plastic sheet spread inside the excavation. To avoid sunlight degradation of plastic sheet, the parts above water level should be covered with a thin layer of soil. The sheet is simply laid into the hole, provided that no stones or other sharp protrusions exist. It is secured, after the excavation has been filled with water, by digging its edges into a trench around the bank.

Both types of reservoir lining need careful treatment during construction and use to prevent puncture.

5.2.2 Water distribution

In the U.K., most irrigation water is carried in pipes, rather than open channels. Piping can be divided into two types, permanent underground and portable above-ground.

5.2.2.1 Permanent underground systems

Piping materials used are as follows.

(a) Rigid plastics, normally unplasticised PVC (UPVC) or ABS, with the joints being made by 'solvent welding'. The end of the pipe fits tightly into the socket on the fitting, after a volatile solvent has been applied to the joint surfaces to melt the plastic in the joint area, so causing the two surfaces to fuse together. Rigid plastic pipe is easily handled and cut, offers low resistance to water flow, is resistant to water corrosion and can accommodate soil movement without breaking. Solvent-welded joints must be made properly at the first attempt since leaks cannot be rectified by dismantling and remaking the joint.

(b) Concrete, either reinforced with asbestos fibres (asbestos cement), or reinforced with steel and compacted by centrifugal force (spun concrete), in which simple spigot and socket joints are often used. The pipe end is a sliding fit in the socket; a rubber ring inside the socket forms the water seal. This does not hold the joint together, but the parts are prevented from moving apart by soil friction when the pipe is buried.

Concrete pipes are relatively cheap, but heavier and more difficult to cut than plastic. Concrete is resistant to corrosion by water, and the loose spigot joint will allow some movement in the soil. The resistance to water flow depends on the internal surface finish.

(c) Metal – both iron and aluminium have been used. Larger mains are made in cast iron, with socket joints as for concrete. Small mains, up to about 50 mm diameter, use galvanised steel water pipe with screwed joints. In most soil conditions, iron and galvanised steel pipe will have a life of many years. Screwed steel pipe is simple to instal and systems can readily be altered with the use of wrenches. Although steel is suitable for welding, this should be avoided on galvanised pipe. The heat destroys the protective zinc finish, leading to rapid corrosion around the welded area, unless it has been specially treated.

Aluminium is light to handle during installation and is resistant to corrosion. The pipe is normally thin walled, so cannot be screwed for jointing. Sockets and spigot joints are used instead.

(d) Flexible plastic. This is used for small mains up to 80 mm. It is easy to instal and does not corrode. Smaller diameters can be installed by 'trenchless' methods, using a mole plough with a pipe feed tube fitted behind the leg. Jointing is made with 'compression' fittings in which a soft copper ring (olive) is squeezed around the pipe wall as the joint is tightened. A metal sleeve (ferrule) is fitted inside the pipe to prevent its wall being deformed inwards by the olive; the ferrule size is dependent on the pipe wall class (thickness). These joints can be dismantled and re-assembled, requiring only a new copper ring. The joints are relatively expensive, but the pipe can be obtained in coils up to 100 m, so reducing the number of couplings needed.

The pipe is obtainable in several wall thicknesses and strengths, classed alphabetically A to E, A being the thinnest. Most irrigation systems are laid in class C, which will accommodate a water pressure of 900 kPa at ambient temperatures.

5.2.2.2 Hydrants

These are the points at which a branch of the main is brought to the surface to supply water to the irrigator. The hydrant pipe diameter is often less than the main pipe where water flow rates into the branch are only a portion of that flowing in the main.

The hydrant pipe should have a blanking cap in addition to a hand-operated valve. The valve will enable the branch pipe to be fitted without shutting off the main supply. The cap prevents localised flooding when the irrigator is disconnected, should the valve become worn or not shut completely.

The hydrant is normally recessed below ground level to allow vehicles to drive over without being damaged. It should be within a brick or concrete chamber, fitted with a lid strong enough to carry a vehicle.

On sports fields the hydrant chamber can be placed adjacent to pitches, with lids at grass level to enable mowers to pass over them. In field-scale horticulture, the mains normally follow the field edges or roadways, with the hydrants at convenient spacings. If the hydrant is on the cultivated area, it should be made prominent, to ensure that operators of cultivation machinery can avoid it.

5.2.2.3 Portable mains

These are used to carry water from the hydrant or independent pump to the irrigation equipment.

(a) For small systems using up to 50 mm diameter pipe, flexible plastic can be used. This might be of the same sort as described earlier, or a more flexible type of woven fabric reinforced light plastic. Connections can be made by screw ring couplings, which can be made watertight using finger pressure only, or bayonet couplings, which lock together when turned through 90°. An alternative is to clamp the hose end over a stub of steel pipe using a worm-drive clip.

Portable mains of flexible pipe will allow the irrigation equipment to be moved around without disconnecting.

(b) Aluminium pipe is usually used for larger portable mains. These normally run from 50 mm to 150 mm diameter. The pipe is thin walled and the standard lengths of 10 m can easily be handled by one man. The pipes are jointed by special sockets which allow rapid coupling, the two common types being either a rubber-lined socket into which the next pipe end is pushed and retained by a latch, or a spherical socket and spigot held together by over-centre clamps. Both types allow for some angular displacement, so that the pipeline stays watertight even when the run is not laid straight. This allows for some flexibility when running the pipeline to the irrigation equipment, but serious changes in alignment have to be taken up with short lengths of flexible pipe.

5.3 PUMPS

5.3.1 Pump characteristics

Most systems use a pump to provide water at the required volume and pressure. Most pumps are of the centrifugal type, where energy is imparted to the water by centrifugal force as it passes through a spinning rotor. The flow of water that the rotor can generate is governed by the ease with which it can leave the pump. The higher the resistance against the water as it flows through the pump, the less easily will it pass through; thus in a centrifugal pump the volume will be proportional to the pressure it operates against. The characteristics of a centrifugal pump, in terms of water volume flow and pressure, are commonly represented in graphical form. Often a pump manufacturer will produce a 'family' of models to the same design, but differing in size or operating speed. Their characteristics are often combined on one graph, with a different line for each pump size (figure 5.3).

This makes selection easier, as it is simple to pick the best sized pump to give the required flow and pressure. Often, the 'size' of a pump will be quoted, for instance a '65 mm pump'. This refers to the nominal pipe diameter of the inlet; thus the pump will have a 65 mm diameter inlet.

Volume flow through a pump varies with impeller speed. In a directly driven electric unit, this is fixed by the motor speed. Where it is driven by an engine or tractor, the running speed can be varied to advantage where a big pump set can be slowed to reduce water flow when operating only a small number of sprinklers.

There is a limit to the water pressure that a centrifugal pump can develop. As pressure increases, the flow will decrease until it ceases. At

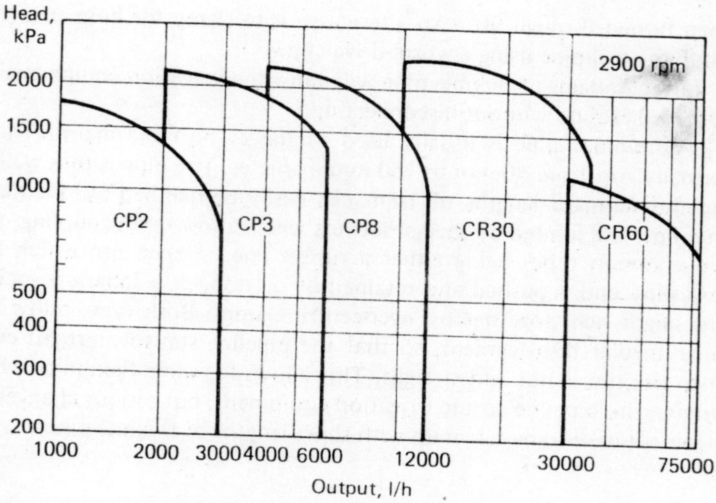

Figure 5.3 Typical pump characteristic chart.

this point, the pump is said to be 'stalled'. This refers to the stopping of the water flow, not to the impeller, which will continue to spin at its normal speed. To increase the pressure characteristics of a pump, the technique known as multi-staging is used. This involves coupling two or more centrifugal pumps together, so that the water flows through each in turn. Each stage adds its own pressure to the water, so that a two-stage pump theoretically produces twice the water pressure of a single stage, a three stage, three times and so on. The pressure requirements of most simple irrigation layouts come within the capabilities of single-stage pumps, but long main lengths and some mobile units require the higher pressures from a two-stage pump.

5.3.1.1 Suction requirements

Although the delivery side of a pump can produce high pressures, the inlet pressure or 'suction head' that a pump can exert is limited by natural laws. Theoretically this is 100 Pa but, in practice, pump performance will be impaired if the suction head is over 60 kPa. In practice, this means that a pump cannot suck from a water surface much greater than 6 m below it. Where conditions dictate that the water source is at a low level, for example in a deep stream, the pump will have to be lowered to within 6 m of the water level. In the case of boreholes, where water may be over 100 m below the surface, the pump has to be placed down the bore. Special pumps are available for this, some of which have the impeller at the base of the bore, and the motor at ground level, driving through a long shaft.

5.3.1.2 Priming
A centrifugal pump creates its flow by the centrifugal action of the water passing through the impeller. If the pump is dry, there will be no water for the impeller to act upon to create the necessary suction, and the pump will run, but without moving any water. To overcome this, water must be introduced into the impeller at the time of starting. Two methods exist.

(a) Use of a small piston type hand pump, arranged to draw water up the suction pipe and through the impeller.
(b) Placing the pump below water level, commonly termed 'flooded suction'.

The alternative of fitting a non-return valve to prevent the suction line emptying is not wholly reliable, as it may not seal completely, although it does make priming and running easier.

5.3.2 Pump seals
Running without water can prove detrimental to most pumps. The impeller shaft runs through a seal in the casing, to prevent water loss. Most seals are simply made of flexible packing material, which is not ideally suited for running against the steel of the shaft. To prevent rapid wear, a thin film of water has to pass between the seal and the shaft, to act as a lubricant. This is the reason that a pump seal will always drip slightly.

Some pumps are fitted with 'mechanical seals', made of materials that can run together without water lubrication. These are more costly than ordinary packing, and tend to be used only where dry running might be a problem.

5.4 APPLICATION EQUIPMENT
This falls into four basic types: static sprinkler, oscillating line, mobile and drip.

5.4.1 Static sprinklers

5.4.1.1 Types available
As the name suggests, these remain in one position and irrigate the surrounding area. The means by which water is distributed are many; the most common types are outlined below.

(a) The static hydraulic nozzle, which is similar to a crop spray nozzle, is installed to spray upwards, and produces a fan or cone of fine droplets which fall on to the crops. The droplet size and spectrum are not as critical as for crop spraying, so this nozzle can be formed from moulded plastic rather than being precision engineered.

(b) The pin nozzle consists of a single-hole jet discharging on to a conical deflector, and fixed on to the head of a pin protruding from the jet hole. The clearance between the pin and the jet hole, together with the deflector shape, govern the spread pattern and water output. The nozzle head and pin are normally colour-coded. each colour relating to an output rate and droplet spread diameter.

(c) In the anvil jet, a high-pressure jet of water, created by a nozzle with a cylindrical bore, hits a metal disc (anvil) a few millimetres away. The anvil can be flat or cone shaped, with its apex pointing towards the nozzle. The water jet is deflected into a circular spray of droplets by the anvil. This jet is always used with the anvil horizontal, to produce a lateral droplet discharge, but may be positioned with the nozzle discharging upwards or downwards.

The spread and droplet pattern is governed by the velocity of the water stream leaving the nozzle, not the clearance between it and the anvil.

(d) In the spinning deflector, the stream of water from a jet is directed on to a specially shaped deflector which is free to rotate. The stream is deflected sideways, but the reaction of its hitting the deflector rotates it. The resulting circular watering pattern is caused by a horizontal stream of water moving around the sprinkler, not a circular spray of droplets.

(e) The rotating head sprinkler also distributes water by moving a horizontal stream around in a circle. The water stream is created by an outward pointing jet, mounted into an assembly rotated by water power. The rotation force is created by a horizontally pivoted pendulum arm which intercepts the water from the jet. The force of the water pushes the arm sideways against a spring which returns the arm to its original position, where it comes into contact with a stop. There is sufficient weight in the pendulum for its momentum to knock the nozzle head round by a small amount as it hits the stop. Pendulum geometry and its spring pressure are designed to keep it oscillating at a regular rate, thus rotating the sprinkler in a series of pulses. The pendulum breaks the water trajectory each time it returns. This helps deflect some water close to the sprinkler, and provides a more even distribution pattern. As the pendulum obstruction tends to limit water trajectory, larger sprinklers have another nozzle on the opposite side of the head. This is not impeded by the pendulum deflector, and so can propel its jet farther to water a wider area.

The rotational drive for some larger sprinklers is taken from a small turbine placed in front of the nozzle. This drives a bevel gear which engages a gear ring placed around the base of the sprinkler.

(f) Rotary wing sprinklers have a jet at each end of a freely rotating arm. The jet angle is such that the force of the water jet leaving it causes the arm to rotate.

5.4.1.2 Discharge characteristics and uses

Static sprinkler discharge rates and effective watering distance depend somewhat on the water pressure and nozzle rating; table 5.6 shows the likely range for each type.

The sprinkler type can determine its usefulness in certain crop situations, the most important being as follows.

Table 5.6

Nozzle characteristics

Sprinkler	*Hydraulic*	*Pin*	*Anvil*	*Spinning deflector*	*Rotating head sprinkler*	*Rotary wing*
Working pressure (kPa)	75-200	100-250	75-200	100-200	200-500	100-150
Water flow (litre/min)	0.5-2	3-7	5-12	1-3	10-200	1.5-30
Effective diameter (m)	up to 1.5	3-6	3-6	2-5	5-25	4-15

(a) Crop height – crops up to around 2.5 m high require high level sprinklers with an upwards trajectory to distribute water above the crop canopy. Tree crops would require an impractical height, and can be satisfactorily watered by low discharge ground-level sprinklers.

(b) Crop area – if the sprinkler covers more than the crop area, it both wastes water and might be disadvantageous to surrounding crops or activities. Large areas, such as sports fields, can benefit from wide trajectory units, especially when fixed to permanently plumbed hydrants.

(c) Crop tolerance – some seedlings cannot tolerate large droplets. Uneven application can result in some parts receiving too much water, which can wash away compost or nutrients.

5.4.1.3 Layout considerations

The area watered by any of the foregoing sprinklers is, theoretically, a circle centred on the sprinkler. The distribution across the circle is not uniform and, in most types, tends to diminish gradually towards the outside, so that the area receiving enough water will be less than the area of the wetted circle. A further problem arises in that most crops are set out in straight-sided areas. Therefore, for a single circular sprinkler to water the corners of a square plot, the central portion will receive too much water. In some crops this will be merely wasteful, but often it will also cause waterlogging or nutrient leaching of the central area.

For most crop situations the sprinklers have to be placed in multiples to cover the desired area. The simplest of these is a single fixed line, and is used by many nursery stock growers and garden centres to water runs of standing ground, or open frame beds. Sprinklers are mounted along a supply pipe at set intervals, which are less than half the wetted diameter, so that the discharge from one complements those on either side. This can produce an even distribution along the line, but there will still be loss of water to either side of the bed area. Where this is undesirable or significantly wasteful, most sprinkler types can be obtained in a form that waters only a part circle (sectoring). On the hydraulic and anvil types, the modified discharge pattern is achieved by shaping the orifice or anvil, normally limited to 180° (half circle), but this allows them to be placed along either side of a bed, with the spray pattern pointing inwards. The rotating head type can be obtained with a 'sectoring' device in which the nozzle rotates against a spiral spring; at a set point in the rotation circle the pendulum action is reversed and the sprinkler is returned rapidly to its starting point.

A sectoring sprinkler can be adjusted for spray arc. These are useful in areas with public access, such as garden centres and sports fields, where one area can be watered without affecting people on an adjacent area.

5.4.1.4 Supply pipework

Supply pipes to sprinklers are normally used to support them as well. Each run of pipe carries one or more sprinklers, depending on their operating centre distance.

Small jets, such as pin nozzle and anvil types, are fixed directly into the pipe wall. Where plastic pipe is used, the thread is formed directly in the pipe wall; on thin-walled aluminium pipe, special threaded bosses are first riveted into the wall. On layouts where more than one pipe length is used, the sprinklers on each section have to be aligned to spray in the same direction. Where fixed pipelines are used, this is achieved when assembling but, on portable systems, it has to be achieved by coupling, which can only be assembled in one plane and usually involves a bayonet coupling or a socket and spigot with an alignment peg.

One sports-field system employs a series of single-hole nozzles fitted in clusters on the pipe. The nozzles forming each cluster point in different directions, so that a lateral spread is achieved by a number of jets of water spraying outwards from the stationary pipe. The pipe stands are in the form of wheels, with the pipe at the centre, so that the pipeline can be moved by rolling, rather than needing to be carried.

Rotary head sprinklers are normally mounted on the top of a standpipe, which rises from the ground-level supply lateral. In most portable systems, the designed sprinkler spacing of 6 m or 10 m allows for one standpipe to be permanently fitted to each 6 m or 10 m length of lateral supply pipe, so there is a minimum number of parts to be coupled together when assembling a sprinkler run.

The simpler sprinkler systems require the entire lateral layout to be carried to each new setting position. This is time-consuming and can be unpleasant work for the staff who have to walk through wet foliage. Two methods have been adopted to lessen the amount of lateral movement associated with resetting; these are known as 'alternate sprinkler positioning' and 'semi-permanent solid set'.

(a) The alternate system uses laterals fitted with self-sealing tee branches in place of the standpipe. A separate standpipe is used which locks into the branch and opens its valve. In use, the laterals are laid out every other day in longer lengths than for normal lines. The standpipes are placed in alternate branches for one day, then moved into the other ones for the next day.

(b) The semi-permanent solid set uses the same laterals as the alternate, but sufficient are laid to cover the whole cropping area at the beginning of the irrigation season. The number of push-in standpipes equating to a single setting is used, and these are moved to new positions to suit the irrigation frequency requirement. The lateral layout remains until harvest.

Both systems reduce labour requirements, and the semi-permanent system reduces crop damage caused by dragging pipes. The systems are more expensive than a simpler fully portable setting.

In orchards a fully permanent lateral system can be installed, where the laterals remain for the life of the trees. To facilitate mowing, the laterals and branches can be placed below ground level.

5.4.2 Oscillating sprinklers

These apply water to a rectangular area by moving a line of water droplets from side to side.

The spray pattern is produced by a row of simple, single-holed nozzles mounted along a horizontal pipe. The pipe is rolled back and forth, through a fixed arc, to distribute this spray laterally. Pipe length can vary between a 'lawn sized' unit, 250 mm long, to a field version of over 50 metres.

Power for oscillation is provided by the water supply. The simplest method involves a water tank on an arm, which slowly fills with water and then empties. As the tank fills, its weight increases and it pulls the arm downwards, causing the line to rotate. At the bottom of its travel, the water supply is cut off, and the tank allowed to drain slowly. As the weight decreases, a spring pulls it back to its starting position, thus rotating the line in the opposite direction. The oscillation rate of this type is very slow, taking about 30 minutes to complete the cycle.

Other systems involve one or two pistons acted on by the incoming water pressure. These fill and empty in sequence, under the control of a valve which is switched at each extremity of the rotation. These offer a much faster oscillation rate, 2–5 minutes per cycle being common. On some machines the switching trips can be adjusted to alter the rotational arc, and hence the spread width.

Powering the sytem using the supply water absorbs a small amount of the available flow or pressure. If the power system is in series with the line so that all the water has to pass through it, some loss of pressure at the nozzles will be experienced. If the power unit is of the bypass type, some flow will be diverted from the nozzle supply, but the pressure of the main flow will not be affected. In this type, the water consumed by the pistons cannot be passed back into the line and drains away. Normally the balance tank type can tolerate a lower water pressure than the piston type.

For field-scale operation, the layout is constructed from sections of pipe, coupled to an oscillator unit. The pipes have quick-fit couplings for ease of assembly, which contain a locating spigot to keep the nozzles on each section in line. The pipes are normally 4.5 m long, and 25–50 mm diameter. The practical length of line that can be used will depend on pipe size and the nozzle flow rates. As a guide, a 25 mm line can have a maximum run of 90 m and a 38 mm line a maximum run of 200 m.

The pipe run is supported above crop level on small stands at 4.5 m intervals. The stand has a horizontal top with curved ends to allow the pipe to roll back and forth as it is oscillated, without the need for bearings. The curved ends prevent the pipe from rolling off the stand.

5.4.3 Mobile irrigators

To avoid the need for constant resetting when irrigating large areas, it is possible to mount the sprinkler system on a carriage, and to move it slowly across the ground. The mobile unit normally requires manual positioning once per day, and moves automatically thereafter. Small machines for market-garden and sports-field use can irrigate 0.4 ha (1 acre) to 1.2 ha (3 acres) from one initial positioning. Large machines for field-scale crops can cover 10 ha (25 acres) or more.

Power to move the irrigator is derived from the water supply, a variety of propulsion systems being used.

The water application rate is a function of the irrigator's travel speed; the faster the machine moves, the lower the rate. Travel speed or 'winding in speed' is variable on all machines, typically ranging between 0.15 and 1.9 m/min to give precipitations between 12 mm/h and 2 mm/h.

5.4.3.1 Hose reel

Integral hose reel units have the propulsion motor, sprinkler assembly, and a large reel containing the water supply hose, carried on the mobile carriage. On most integral machines the propulsion motor acts on a winch wire, instead of pulling on the hose. The winch wire is fixed to a sprag, and the machine drawn towards it. The water-supply hose is flexible plastic, from 50–125 mm bore, depending on the machine application rate. On some irrigators the hose reel is designed to unroll only, requiring the carriage to always work away from the water main. More sophisticated types have a driven hose reel, so that the hose can follow the winch wire and be rolled up as the irrigator progresses. On most irrigators of the latter type the run can be twice the hose length, this being achieved by placing the supply main at the halfway point so that the hose is rolled up as it approaches the main. When the main is reached, a trip disengages the hose reel drive, so allowing the pipe to be paid out.

A few machines do not use hose reels, the hose being left to trail across the ground surface. These are suitable only for crops that the hose cannot spoil by dragging over, such as amenity grass.

5.4.3.2 Towed carriage

In this type, the hose reel and winch unit are stationary during use, and pull a light carriage across the field which holds only the sprinkler. The carriage is easier to pull out along the crop rows when setting up, and produces less ground pressure and soil compaction. The run length is restricted to one hose length, but the static reel can often carry more hose. The carriage is often too light to carry more than one sprinkler nozzle.

Both types of mobile irrigator are designed to spread water behind only; this avoids waterlogging the ground ahead of its wheels.

In stony ground, machines that drag hose along the ground are not popular owing to high rates of hose wear from stone cuts. The travelling reel type is preferred, as the hose is laid down and not dragged.

5.4.3.3 Propulsion units

Propulsion units fall into two distinct types.

(a) The turbine system involves a rotor, which is placed in the inlet pipework so that the flow turns it at high speed, and a series of gears or pulleys is used to convert this motion to the reel rotation speed. Often a variable V pulley is included in this transmission to enable the reeling speed to be varied. The turbine adds to the pipework flow resistance, and the turbine machine normally requires a pressure 100 kPa (15 psi) higher than the piston type.
(b) The bellows or piston type uses the pressure of the water to oscillate a ram or rubber bellows, which drives the reel through a ratchet system. The ratchet system often acts directly on the hose reel rim, so obviating the need for gearing. On machines using double-acting rams, the ratchet system is also designed to drive on both strokes. Oscillation is controlled by a trip valve which diverts water into the appropriate end of the ram, or causes the bellows to fill and drain alternately. The water used is fed from the main supply, but runs to waste afterwards. Thus, while some flow is lost, the remainder does not suffer any pressure drop.

The small amount of water drained from the motor is discharged on to the ground beneath, and can cause localised waterlogging if it is on a static reel machine. Travel speed is regulated by adjusting the useful travel on the ratchet drive.

Water quality is normally more important to the piston or bellows type, as grit can affect valve seating, and cause wear on sliding parts.

The units contain a trip device to stop water flow at the end of a run and also their travel. On many systems the mains pressure surge caused by the trip valve activates a pump cut-off.

The water motor provides slow speed winding for use when irrigating. Rapid reel winding when making the machine ready to move to a new position is accomplished by a PTO or engine drive direct to the reel.

The reel speed has to be varied during the winding process, to take account of the increase in diameter as the hose is wound on. Without this the carriage would increase in speed as it approached the reel; for example, if the reel is rotating at 1 rpm, the carriage will travel at 6.3 m/s when the hose reel is 2 m diameter and 9.4 m/s when it is 3 m diameter. The effective winding diameter is sensed by a roller running against the hose on the reel, which alters the piston stroke or turbine speed.

The large-bore hose of high output mobiles contains a large quantity of water. This adds weight which can make the hose reel unit difficult to move. It also makes the hose too stiff for rapid winding or unwinding when the machine is being repositioned. There is thus a separate facility to drain this water, the most common being a small low-pressure air compressor, driven by the tractor PTO, to blow the hose clear.

5.4.4 Drip feed systems

These are designed to apply water to the plant root zone, as opposed to an overhead spray. The system offers a much better use of water in widely spaced plantings. Three systems are used.

(a) Pipes laid on the soil surface or across the tops of containers. Water is discharged either directly by drip nozzles fitted into the pipe wall, or by micro-bore lateral tubes. The former system is only useful for single rows of plants in a relatively straight line. Micro-bore laterals can accommodate groups of plants, such as containers on a standing ground, with one lateral running to each. In some micro-bore systems the resistance to flow is sufficiently predictable to enable tube length to be equated to water output.

(b) Sub-surface 'drip' lines are used in orchards or with other long-term crops. The main advantage in burying the pipeline is to facilitate mowing and weed control. The pipeline is flexible plastic, and can be laid by moling, either before planting or shortly afterwards, but before full root development has occurred.

The water outlets are normally positioned above ground level to enable a visual check on their performance to be made. The outlet will be either a vertical pipe, if the main pipeline runs beneath the roots, or terminated in a flexible tube which can be laid on to the root zone if the main pipeline is to one side.

(c) Seep hose is used where the crop rows have such a high population of plants that it is impractical to water each one individually. The most common type is a thin-walled sheet plastic strip, which is folded over into a tube and the joint loosely sewn to allow water to seep out. As this will have a limited operating length before the water drains out, systems have been devised to refeed it at intervals; one such method is a seep tube bonded to a water-carrying tube, with connecting holes at intervals to resupply the seep hose.

Seep hose is normally laid on the surface, although it has been moled in beneath row crops in arid areas to provide water without any loss from evaporation. Seep systems are affected by uneven soil surfaces, since most water is applied in the hollows. A dirty water supply or the presence of algal growth causes the water escape areas slowly to become blocked.

Most drip systems operate on low pressure at 20–100 kPa, and normally only sufficient head is required to push the water through the seep holes. It is usually sufficient for the water supply to come from a tank placed a few metres above pipe level, rather than to require pumping or to be under direct mains pressure.

Note that nutrient diluters are covered in the companion volume, which deals with fixed equipment.

5.5 EQUIPMENT SELECTION AND OPERATION

5.5.1 Selection

When selecting the equipment for a system, two aspects must be considered: soil infiltration rate, and droplet size.

(a) Infiltration rate governs the rate at which water can safely be applied. The water must use the soil pore spaces in order to travel from the surface into the root zone. The pore size is greater in a sand or good loam than in a clay. Water swells the soil particles so that, following the initial wetting, the pore space will reduce and the infiltration rate will slow. Water that is applied in excess of infiltration rate will stay on the surface and, if on a sloping site, will lead to run-off erosion. The maximum infiltration rate for a soil is difficult to quantify, but a useful 'rule of thumb' is to limit the field application rate to 12 mm per hour. Most commercial field equipment systems are designed to apply this rate or less.

(b) Droplet size is critical for soil and plant damage. Very large droplets can cause the surface particles of some soils to compact together, so forming into a hard cap when dry which can restrict the growth of small seedlings. Small, delicate plants can be damaged by large droplets landing on them, and soil particles splashed up by large drops can contaminate low-growing salad crops like lettuce. Droplet size is related to the nozzle orifice diameter of the sprinkler. Very large sprinklers (commonly called rainguns) can have a single nozzle of up to 25 mm diameter, which will produce drops capable of causing damage. It is recommended that large sprinklers and rainguns be limited to a nozzle diameter of 14 mm for all horticultural crops, except established grass.

Very fine drops cannot be projected far, and are easily carried by the wind. This means that fine misting nozzles, like the pin and anvil types, are unsuitable for field-scale operation.

5.5.2 Operational checks

(i) The main check must be on evenness of application. For multiple-nozzle or sprinkler systems, pipe resistance will result in a gradual diminution of the output of nozzles, as their distance from the input increases. The length of pipe at which this becomes critical will depend on its diameter and the flow of water. Standing to one side of the line, at a distance, when it is working will reveal any significant problem, as the height of the water spray will vary. If this becomes visible, it should be cured by reducing the line length, substituting the first few runs for larger pipe, or feeding into the centre of the run.

The application evenness can be checked quantitatively using catching cans. These are a number of open-topped containers of the same size – $2\frac{1}{2}$ litre paint tins are ideal – set out in the irrigated area. The depth of water in each can is measured and, if these are plotted on a graph against the can setting spacings, the distribution curve will become apparent. In addition to faulty nozzles, the can will show any variation due to wind, and where nozzle spacings are too wide or too narrow. The depth in each catching can also give a direct indication of the rainfall equivalent in mm or inches.

(ii) Droplet size measurement is more involved. The most usual method is to expose, for a few seconds, a shallow tray of flour to the falling droplets. This involves working in the irrigated area. The water in each drop causes flour particles to coalesce in small lumps. The tray is placed in a hot oven to harden these lumps, and the contents sieved to remove the loose flour, so that the hardened 'drops' can be measured and counted.

(iii) Normal maintenance will include checking free operation of all rotating items on sprinklers, and cleaning silt and algal growth from the nozzles. Pipe coupling seals must be in good order to prevent localised ponding from leaking joints.

(iv) Mobile irrigators must be checked for correct tracking shortly after commencing a run, as running off-line can destroy crop rows beneath the machine. These machines should not be used when there is a strong tail wind, as water blown forwards can soften the soil over which it travels.

5.6 CALCULATION STEPS FOR A STATIC SYSTEM

Using some or all of the following steps, a complete irrigation system can be designed and equipment specified.

5.6.1 Quantity of water (total per year)

Total the quantity from each crop as in section 5.1.1 to give the total reservoir or annual extraction requirement.

5.6.2 Reservoir design

Calculate the required capacity as in section 5.1.3.

5.6.3 Area to be irrigated at each setting

For each crop: (i) divide the total area to be irrigated by its minimum recommended frequency in days; (ii) total the setting area requirements for each crop, and find the maximum area to be covered at one setting.

If the recommended precipitation rate is such that it can be applied in less than 8–10 hours, it might be possible to cover two settings per day; this will halve the setting area requirement for that crop.

5.6.4 Amount of equipment

The manufacturer's data sheets are used to establish the layout; an example of one is shown in table 5.7. It shows the range of nozzle sizes that can be fitted and the water flow at various pressures, and on the right-hand side it shows the precipitation rates at a range of spacings. It is usual first to choose a suitable grid spacing, which will normally be governed by pipe length increments of the main and sprinkler pipe, and then to find the nozzle size and pressure that provide the required precipitation rate.

Table 5.7

Sprinkler performance

Nozzle size	*Pressure at nozzle*	*Discharge*	*Wetted diameter*	*Sprinkler spacing in metres to produce the precipitation rate in mm/h*			
(mm)	*(kPa)*	*(l/s)*	*(m)*	*6 X 12*	*10 X 10*	*10 X 12*	*10 X 15*
3	200	0.16	23.0	7.5	7.0	5.0	–
	275	0.18	23.5	8.5	8.0	6.0	–
	350	0.20	24.5	9.5	8.5	6.5	–
3.5	200	0.19	23.5	9.5	8.5	6.5	–
	275	0.23	24.5	10.5	9.5	7.0	–
	350	0.25	25.0	12.0	10.5	8.5	6.0
4	200	0.25	25.0	12.0	10.5	8.5	6.0
	275	0.29	25.5	13.5	12.5	9.5	7.0
	350	0.32	26.0	15.5	13.5	10.5	8.5
4.4	200	0.29	25.5	13.5	12.5	9.5	7.0
	275	0.34	26.0	16.0	14.5	10.5	9.0
	300	0.38	27.0	18.0	16.0	12.5	9.5
	400	0.41	27.5	19.5	18.0	13.0	10.0

If the setting area requirement is greater than individual crop plot areas, two or more complete sets of equipment might be needed to enable more than one crop area to be watered at once.

5.6.5 Flow of water to be supplied

This should be calculated from the total flow from all the sprinklers in the setting above, and compared with the crop calculated flow rate in section 5.1.3. If the two are widely variant, recheck the setting equipment. If the figures largely agree, this will be the flow that the pump must produce and the main pipes carry.

5.6.6 Calculation of the pipe sizes

Using the head loss calculations in section 5.1.4, calculate the pipe friction head. Use pipe flow charts to arrive at the size of each section of main and lateral.

5.6.7 Calculation of the pump head

Pump head is calculated by adding together the pipe friction head calculated in section 5.6.6, the rise in height between the reservoir and the sprinkler, and the pressure needed to operate the equipment.

$$\text{pump head (Pa)} = \text{pipe friction (Pa)} + \text{lift (m)} \times 9.7 + \text{equipment pressure (Pa)}$$

5.7 EXAMPLE OF LAYOUT CALCULATION

Two blocks of lettuce, each of 0.5 ha, are to be irrigated with 44 mm of water over 4 hours. The length of the underground main from the pump to the first hydrant is 150 m, with a further 150 m to the next. Both blocks start 150 m from their hydrant.

5.7.1 Water volume flow

Using the relationships given in sections 5.1.1 and 5.1.2, the volume flow rate is calculated as follows:

Volume to give 44 mm of water on two 0.5 ha blocks $= 2 \times 0.5 \times 44 \times 10\,000$ litres

Rate per hour to give this volume flow in four hours $= \dfrac{2 \times 0.5 \times 44 \times 10\,000}{4}$ l/h

$= 110\,000$ l/h

or $= 30.5$ litres/second

5.7.2 Sprinkler layout

Taking one block of lettuce, 0.5 ha can be represented by a rectangle 100 m x 50 m. The output from each sprinkler must apply 44/4 = 11 mm/h. From table 5.7, it is possible to get 10.5 mm/h using a 4 mm nozzle at 200 kPa, with a 10 m grid spacing; 11 mm/h should be obtained from this nozzle at about 220 kPa.

5.7.3 Pipework size

At this stage a diagram of the layout can be prepared (figure 5.4); this shows 5 rows of 10 sprinklers each. The initial suggestion for sizes of the pipework is

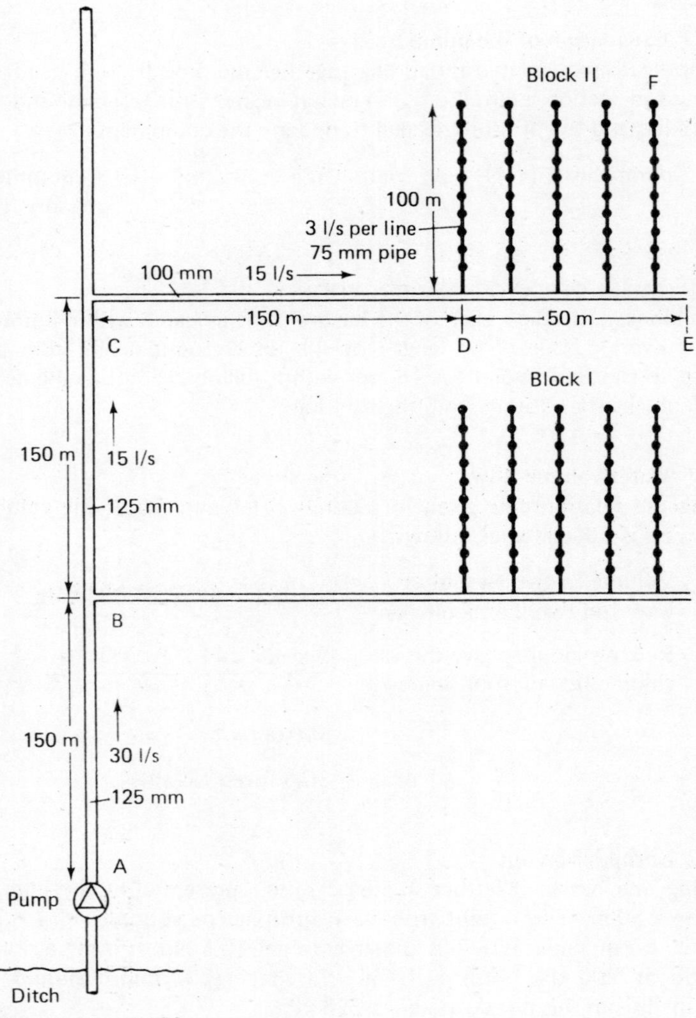

Figure 5.4 Irrigation layout as in example 5.7.

Underground mains	125 mm
Submain to block	100 mm
Sprinkler line	75 mm

Using the pipe friction chart (figure 5.2) and the fittings equivalent length table (table 5.3), the head loss table (table 5.5) can be drawn up.

Run section	*Water flow (l/s)*	*Pipe size (mm)*	*Loss/m (Pa)*	*Length (m)*	*Fitting E.L. (m)*	*Section loss (Pa)*
A–B	30	125	320	150	3	48 960
B–C	15	125	95	150	9	15 105
C–D	15	100	260	150	–	39 000
D–E	15	100	260	50	15	7 436[a]
E–F	3	75	60	100	20	2 272[b]
						Total loss = 113 273

[a]Using m.o.f. for 5 outlets
[b]Using m.o.f. for 10 outlets } (table 5.4).

The pipework resistance for this layout is thus 113 273 Pa or 113.3 kPa, which is not unacceptably high.

5.7.4 Pump size

If it is assumed that there is no rise between the pump and the sprinkler but that the pump is drawing from a ditch 3 m deep, the total pump head will be

113.3 (resistance) + 220 (sprinkler) + 30 (suction lift) = 363.3 kPa (rounding up to 365 kPa)

The pump must thus deliver 30 l/s at 365 kPa.

Note that often there is a slight difference between the pump output as calculated in section 5.7.1, and that calculated by multiplying the number of sprinklers by their quoted flow rates; it is suggested that the higher of the two output figures be used, to ensure that the system meets the required performance.

6 HARVESTING

The culmination of most husbandry operations is harvesting the crops produced. In early times this involved the physical effort of many people but, in more recent times, manual labour has become expensive, leading to the development of mechanical harvesting systems. The obvious aim in harvesting is to remove the crop rapidly from the field, but of equal importance is to remove it in a state fit for market presentation or storage.

6.1 PRE-HARVESTING TREATMENT

Some crops require an operation to prepare them for harvest, ranging from a separate operation carried out some time ahead, to one carried out on the front of the tractor drawing the harvester.

6.1.1 Topping

The most common of these operations is topping; many methods are employed, most being specific to a crop or group of similar crops.

6.1.1.1 Horizontal rotary knife

This is used for removal of tall, light growth. The knife is either a flat bar with sharp tips or a disc fitted with small triangular blades. The knife assembly is often mounted rigidly on the harvester, but for crops like leeks and celery it can float vertically to follow ground contours, and provide a more consistent topping height.

The swede harvester topper is a heavy serrated disc, which can cut through the thickened neck without knocking it over.

6.1.1.2 Reciprocating mower

This is used where the top rubbish is dry but stringy, and will not stand up to impact cutting. The normal single knife mower bar can be blocked easily by large quantities of trash. A better type is the double reciprocating knife 'Busatis' mower in which, instead of the knife blades scissoring against fixed fingers, two sets of reciprocating blades slide across each other. A Busatis mower will tolerate a higher forward speed before blocking, and can cut through growth under strawed-down crops.

6.1.1.3 Flail rotor

These machines have a series of flails fitted to a horizontal shaft. The flail rotor is driven in the opposite direction to forward travel. The machines are capable of dealing with heavy vegetation, the flails being flexibly mounted to allow them to swing back on hitting an obstruction.

As the cut is by impact, the vegetation is shattered rather than cleanly cut, so the machines are more suitable for root crops like potatoes, onions, bulbs and beetroot, where the complete top has to be removed, rather than neatly trimmed. To improve on the quantity of top removed, sets of flails of differing lengths can be fitted, the short one above the crop row to prevent damage, and the longer ones in the spaces between to cut to soil level (figure 6.1).

For onions, where the mature top falls among the bulbs, the air-fanning action of the rotor can be adapted to 'suck' the top up into the

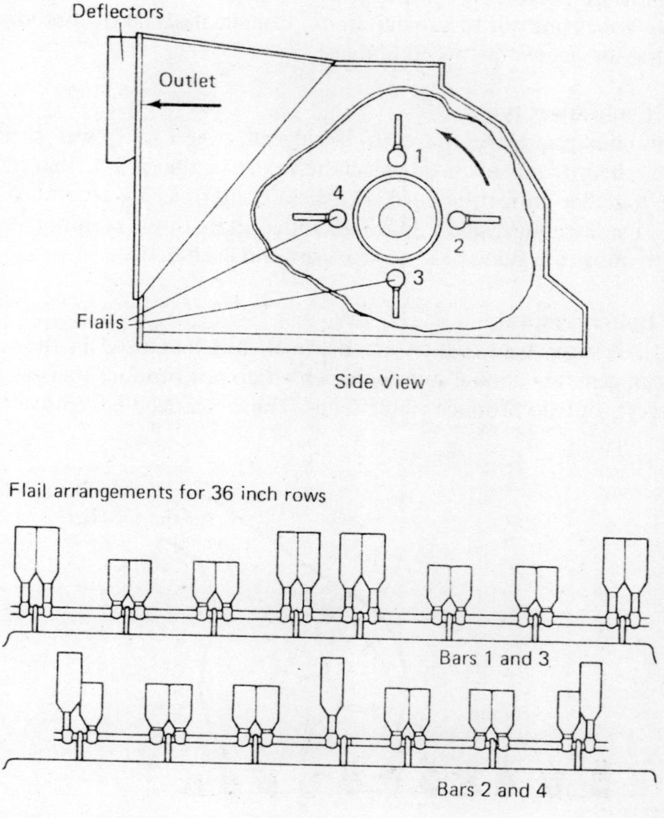

Figure 6.1 Root topper.

flails. This is helped by using flat, wide flails and fitting air deflectors (figure 6.2). Once cut, the flails throw the top out of the machine. The topper might be fitted with an angled chute to throw the efflux upwards and sideways, or might have a transverse auger to convey the efflux sideways so that it drops into the space behind one wheel.

6.1.2 Deleafing

(a) Brussels sprouts. Most mechanical systems are based on poultry-plucking principles; rotors of soft rubber fingers break off leaves without damaging the buttons on the standing crop. The equipment requires space between rows for the rotors to work, and the rotor units have to travel around each stalk. This involves wider crop spacings than normal, which, with modern button strippers tolerating leaves, has virtually stopped further development of this equipment.

(b) Rose bushes. These are deleafed prior to lifting and storing. Most machines use revolving brushes, the bristles being hard enough to remove leaves but not to damage stems. Chemical deleafing methods are reducing the needs for this equipment.

6.1.3 Cauliflower tying

In hot climates, leaves have to be placed over cauliflower curds to protect them from sun scald. Machines exist in the U.S.A. that lift the leaves together above the curd, and tie a bond of string around to hold them. These require plant spacings wider than those common in the U.K. to allow the tying head to work around each plant.

6.1.4 Bolter removal

Crops like sugar beet, which are biennial, but harvested in their first year, can generate annual subspecies which do not produce a viable root at harvest, but do produce viable seeds. There is a need to remove these

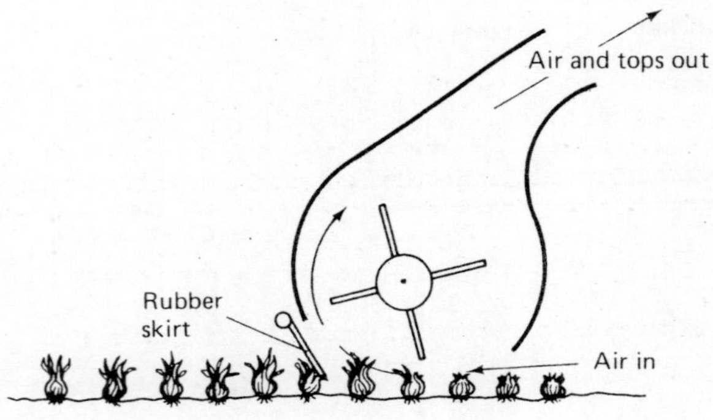

Figure 6.2 Onion topper.

seed heads prior to their ripening. Two methods are used, both relying on the prominence of the seed head.

6.1.4.1 Pulling
Mechanical grippers mounted on a tractor toolbar detect and close on to each seed head, and grip it long enough for the plant to be pulled out.

6.1.4.2 Electrocution
A tractor-driven generator is used to produce a high voltage relative to the soil surface, by means of a metal rod set at a height to touch seed heads. The power runs to earth through the seed head, and is sufficiently strong to kill it.

6.1.5 Tulip heading
The flowers of tulips grown for bulb production have to be removed to prevent disease, but the foliage must be left intact. To ensure that the removed flowers do not fall into the crop, special machines are used in which a mower blade is followed by a collector box, so designed to ingest the cut heads by means of a belt conveyor, paddles or an air blast. The header working height has to be carefully adjusted so that the header cuts all flower heads but runs above the foliage.

6.1.6 Haulm pulling
There is interest in systems for pulling all the top growth off a potato ridge, while leaving the tubers undisturbed and still covered with soil. One principle uses a pair of contra-rotating rubber rollers, which converge over the ridge apex. The haulm is gripped between the rubber surfaces and pulled upwards, but the roller profile and position allows them to press down on the ridge to prevent the soil lifting. Another principle involves gripping the haulm between a pair of inclined rubber belts, with separate means for holding down the ridge top being provided.

Such equipment is the subject of much development, the present problems being stone damage to the rollers, stones and clods getting between the pulling surfaces, so that they cannot close on to the haulm, and the pulling speed, which if too fast can snatch the haulm and break it.

6.2 UNDERCUTTING AND DIGGING

Root crops and plants that have to be lifted complete with root can be hand harvested following mechanical loosening (undercutting) or digging.

6.2.1 Undercutting

A correctly undercut crop is loosened so that it can be easily pulled by hand, but remains upright in the original positions. This can offer simple quality grading, as the labour need pull out only the good plants. Many types of undercutter are available.

6.2.1.1 Bar share

This is a flat steel bar, typically 100 mm wide by 20 mm thick, which is fitted beneath a tractor toolbar, and drawn beneath the crop roots.

The bar is used at a slight angle to break the soil slice upwards, and also to pull it into the soil. To improve the upward heave, a number of steel fingers might be welded on to the back edge of the bar. In hard ground the draught forces can be high, and penetration and depth control difficult. The bar is suitably only for shallow rooting crops like leeks, onions, celery and some nursery stock.

6.2.1.2 Share lifter

When this is used on tap-rooted crops like parsnips and large carrots, the share resembles a pair of plough points, one running each side of the row, which are angled so that the root is squeezed up between them as they pass. Crops with wide and deep roots, where the rootball is to be left intact, like nursery stock trees, require a 'U'-shaped share.

6.2.1.3 Onion undercutter

Crops like onions require to be undercut about 20 mm beneath the surface. A very light machine can be used, often a steerage hoe with one 'L' blade beneath each row. An alternative machine uses dished discs, running nearly flat on to the soil, which are rotated by ground contact to undercut the bulbs and lift them slightly.

6.2.1.4 Vibrating shares

Vibration is used both to lower the share draught force, and to reduce soil adherence to the roots. In most cases the vibration is confined to oscillation of the digging parts, rather than intentionally vibrating the whole machine frame. Three main systems are used.

(a) Vertical vibration to a finger bar following a bar share. On a few machines both bars are oscillated 180° out of phase to one another, to provide a more even load to the drive. Even on machines with an oscillating finger bar only, the vibration caused is transferred to the bar share through the frame.
(b) Fore/aft oscillation. This can be used on a single share, but is more commonly used with a twin bar system like (a).
(c) Lateral oscillation machines. These have been constructed by fixing a bar share beneath each of the tine bars on a two-bar reciprocating

harrow. This latter system provides the best loosening of plant and root soil without causing so much disturbance that the plants fall over.

6.2.2 Digging

The foregoing machines merely loosen, but many root crops like potatoes, beetroot and bulbs can be picked more easily if the crop is left on the surface by a digging machine. This is a combination of an undercutting share with a means of positively raising and separating the roots from the soil. The main methods used are as follows.

6.2.2.1 Plough

This is a modified plough body, in which the main mouldboard is at a lesser angle and made of spaced bars. Used beneath a crop row, the share undercuts it and the furrow slice slides up the mouldboard, so losing loose soil through the bars in the process. Often a twin-bodied plough, resembling a ridger, is preferred as the crop is spread across a wider band, and makes picking easier. A plough is unsuitable for crops where top growth is to be retained, like leeks and nursery stock, as this can be turned under as the roots are ploughed up.

6.2.2.2 Vibrating digger

This combines an undercutting share with an inclined vibrating bar screen. The bar screen shakes soil from the roots and elevates them to surface level. As in the vibrating undercutter, only the screen is intentionally vibrated, but the sympathetic vibration set up in the frame acts on the share to make its passage easier. The discharge from the screen is normally to the rear, but some narrow machines will discharge to the side.

6.2.2.3 Elevator digger

Elevator diggers form the most common type. The share is followed by a powered conveyor of spaced bars, to lift the crop and remove loose soil. Soil removal is improved if the conveyor is agitated by running it over eccentric idler rollers. The conveyor is often broken into two runs, as the drop from the first to the second helps soil removal by re-arranging the mass of material.

Elevator diggers inherently discharge to the rear. Deflectors are fitted to wide machines to restrict the crop wind row width. For side discharge a transverse deflector chute or powered conveyor is added.

Some elevator diggers used for bulbs have both types of soil-sifting system, the first being a run of web conveyor, followed by a vibrating screen. The bar spaces on both the web and the screen are narrow, to retain small bulbs, and the combination has proved to be the best for soil sifting on bulb-growing land.

Most diggers are made to suit potato ridge widths, being either 'single row', 750–900 mm wide, or 'double row', 1500–1800 mm wide. As horticultural growing systems are also based on tractor wheeling widths of 1500 mm or 1800 mm, these machines normally can fit the rows without modification.

6.2.2.4 Spinner

Spinners use a rotating wheel to elevate and clean the crop loosened by the share. Two versions of wheel are used.

(a) A vertical wheel with a series of finger banks around its circumference. These revolve across the furrow slice loosened by a share, throwing it sideways against a vertical screen. The combination of the digging fingers and the screen separate the crop from loose soil. In most machines the fingers are attached to the rotor by a linkage which keeps them vertical.
(b) A spider wheel rotor mounted on a near-vertical axis. The wheel is inclined so that its leading edge runs beneath the furrow slice and its rear end is sufficiently high to discharge the crop on top of the soil. A cage of stationary bars is used to hold the crop on to the spider between these points; the rumbling action between the spider and the cage removes loose soil from the crop.

6.2.3 Selection of machine type

Although the digger might seem a logical improvement on the undercutter, in many cases the crop or its end purpose will dictate which of the two is suitable. The cleaning action of the diggers tumbles the crop into a random mass, rather than an orderly procession. On certain crops the following effects will be noted.

(a) Leeks and celery – the tumbling action allows soil to penetrate into the centre of celery, and the leaf axils of leeks. This soil is difficult to remove by washing. Celery can also be damaged and outer stems lost.
(b) Carrots – the digger is eminently suitable for maincrop carrots, but damages and soils the tops of early crops intended for bunching.
(c) Nursery stock – the cleaning action can remove too much root ball soil, and damage the plant, especially the growing point of subjects like forestry trees, which must develop only single stems.
(d) General packhouse presentation – extra packhouse labour might be required to resort jumbled crops before trimming, grading and packing can occur. An undercut crop can be hand-lifted and field-packed in an orderly manner into crates or bunches, and laid out again with ease at the packhouse reception.

(e) Root crops – which lie entirely below the surface – gain little from undercutting, as the manual workers still have to dig them from the soil.
(f) Shares – the ability of most harvesters to deal with a wide range of root crops is due to the design of the lifting share. Crops like potatoes, onions, beetroot and bulbs require a broad, flat share, while single tap-rooted crops are best lifted by a pair of finger shares, one running each side of the root. Although the main purpose of the share is to loosen and lift the crop on to the harvester intake, it is desirable that it does not lift large volumes of soil in doing so. Therefore flat shares are often only as wide as the crop root zone, and might be slightly hollowed to encourage the material towards the centre rather than spilling off each side. For heavy soil a skeletal share can be used, consisting of a number of flat bars with spaces between. This both allows some soil to drop through and, more importantly, cracks the soil slice sliding over it, so that the slice is more easily broken by the harvester.

In addition to the foregoing 'static' shares, many examples exist of powered or moving shares, the main ones being as follows.

(a) A slightly concave disc running near-horizontal to the soil. The disc is driven and rotates at a speed roughly equal to forward travel, which breaks the soil slice better and reduces tractor draught compared with the static share.
(b) Digger wheels, commonly called 'Oppel wheels', for lifting root crops are thin skeletal wheels, one of which runs either side of the root. They are mounted at an angle so that the gap between them is narrower at their rear, and a root entering them is firmly gripped as they rotate and thus pulled out. The wheels are normally ground-driven, but on some machines one is powered to run slightly faster than ground speed so that the root is also slowly turned as it is lifted.

6.3 COMPLETE HARVESTERS FOR ROOTS

Many common crops can be harvested directly into the trailer or transport container, with little or no manual labour involvement. The machines can be grouped into two types, according to whether the crop is dug or pulled from the soil by its foliage. The digging harvesters can be subdivided into 'manned' with facilities for manual labour to sort the crop from field debris, and 'unmanned' where the crop is cleaned mechanically.

6.3.1 Digging type

6.3.1.1 Manned

The layout of a typical manned harvester is shown in figure 6.3.

The main components are as follows.

(a) Share – this is similar to those used on elevator diggers. On many machines the share depth is controlled either directly or hydraulically, from a roller sensing the ridge top height.
(b) Primary web – the main soil-removal area. This is normally an agitated bar trace conveyor. Clod and soil removal is aided by banks of rubber fingers projecting downwards into the crop as it runs along the conveyor, or by a weighted canvas sheet trailing on the crop.
(c) Haulm extractor – this removes the top growth (haulm) of potatoes and other long vegetative material. three systems are common.

(i) A conveyor with bars about 250 mm apart, which carries long material but allows the crop to drop through. The action can be aided by agitation, or passing the material on the conveyor through a set of 'stripper' rollers which push back the crop.

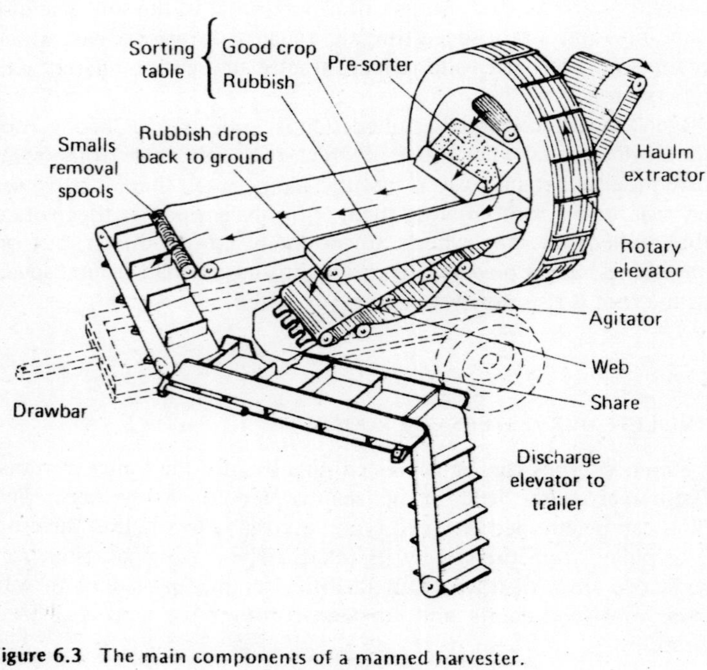

Figure 6.3 The main components of a manned harvester.

(ii) A steeply inclined web conveyor with protrusions to catch hold of long material but allow smaller objects like tubers to roll back. This action is aided by banks of oscillating fingers gently stirring the material on the conveyor.

(iii) A roller placed at the end of each run of web conveyor, and rotating in the opposite direction. The position and size of the roller allows long material to be pulled through, but prevents crop tubers or spherical roots from becoming entrapped. For tap-rooted crops like parsnips and carrots, the roller drive can be disengaged or reversed to prevent their entrapment. Efficiency of removal is increased by fitting curved fingers around the web conveyor end, which guide long material into the roller.

(d) Elevator – to lift the crop to the sorting table. The crop might be carried in pockets around the inside of a large wheel cage, a vertical conveyor with large flights forming pockets, or a series of steeply inclined conveyors. These latter often run with a second conveyor web resting on top, so that the crop is sandwiched between and cannot roll back.

(e) Pre-sorter – to partially divide the crop from stones and clods. Most systems use an inclined flat belt, so that round objects like tubers roll down and flatter stones or clods are carried. The inclination can be varied to deal with some variation in crop shape or soil type. To deal with round stones, a 'hedgehog' belt is used. This has a deep pile of long rubber pintles, which is able to support the crop material but traps heavier objects like stones. The crop is brushed from the belt surface on to the 'good' side of the sorting table.

(f) Sorting table – normally a pair of parallel flat belt conveyors fed from the pre-sorter so that one carries mostly 'good' produce and light debris, the other mainly soil and clods. People can stand either side to divide the crop further from field debris. The soil belt discharges on to the ground, while the crop belt discharges to the elevator.

(g) Discharge elevator – carries the crop to the trailer or other container. To prevent damage from dropping the crop an excessive distance into the trailer, its discharge height can be varied. Many elevators are made in two sections, which can be angled independently into a swan-neck shape to clear the sideboard of a trailer, but still discharge near the floor. On some machines the elevator intake is enlarged into a hopper which holds up to 1 tonne of crop, and allows the harvester to work independently of the trailer taking away the crop.

(h) On some harvesters the trailer elevator can be replaced by a bagging unit and handling platform for crops that are marketed directly from the field.

6.3.1.2 Unmanned harvester

There are three versions of the unmanned machine.

(a) The first type resembles an extended elevator digger, fitted with a transverse elevator at the rear to convey the crop into a trailer (figure 6.4). The digging web is made of three or more sections of conveyor, with a contra-rotating haulm extractor at each conveyor junction. Soil removal is aided by stirring the material as it progresses up the webs, using downwards projecting fingers. A bank of cleaning rollers is placed at the top end of the web run, the rollers being in the form of plain cylinders, star spools, disc spools or heavy coil springs; this helps to break soft clods and generally improve soil removal.

The severity of the soil-extraction action can be varied by the degree of web agitation, the angle of the stirrring fingers and the slope of the final roller bank.

(b) An adapted manned harvester uses short-wave radiation (x-ray or gamma-ray) selection. Radiation of these wavelengths will pass through vegetable matter but not through stones or clods, so these latter items will be detected as 'shadows' in the radiation beam. The selector replaces the normal sorting table and is positioned so that the crop falls past the sensor head towards a bank of sorting fingers. The sensor programs the appropriate finger to deflect the crop items on to the trailer-loading conveyor and to release stones and clods back on to the ground.

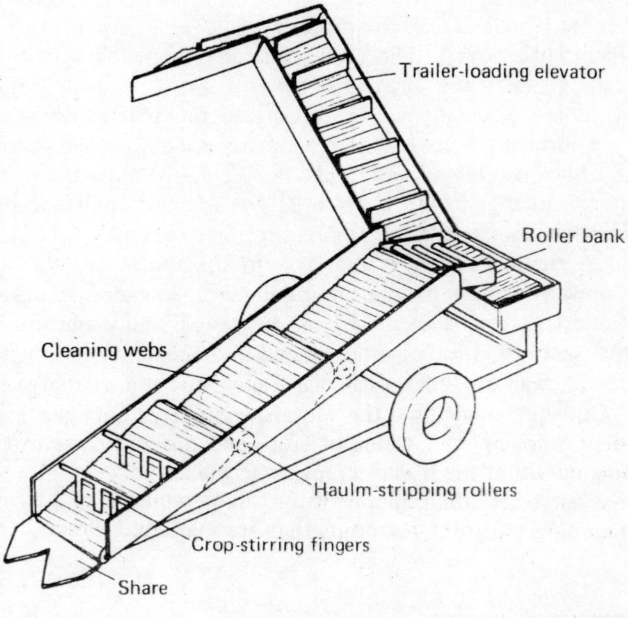

Figure 6.4 Unmanned harvester.

(c) Several attempts have been made to deflect the lighter vegetable matter away from stones or clods, using the force of an air current. In this type of harvester sorter, the harvested material either travels over a gap between two conveyors, where an upwards air current supports the crop across but allows the heavier material to drop, or falls through a horizontal air stream, where the lighter crop matter is deflected sideways. This method of selection is more successful in dry conditions, and it also requires a high power input to create an air current of sufficient velocity.

Bulb harvesters differ from the foregoing types because of special requirements for handling the crops, and the sandy soil on which they are grown. As bulbs multiply by subdivision, each 'root' harvested is a cluster of individual bulbs, held together by root growth. The primary web of a bulb harvester is followed by several rows of rotating star-shaped, soft rubber spools, which pull the clusters apart and also help soil removal from the dense root mass. The crop then passes over an oscillating bar screen which shakes out more soil and very small bulbs; some machines have facilities for hand sorting at this point. The bulbs are discharged into a large bin carried on a frame, which can be tilted to allow them to roll down the inside rather than fall directly.

As bulbs form 75–100 mm below the soil surface, most harvesters are fitted with a ridge plough to bulldoze surplus soil aside. The most common type is an oscillating V plough, driven in opposition to the bar screen to balance machine vibrations. Other types include a powered vertical shaft paddle wheel, or a set of discs arranged in V formation.

6.3.2 Picking up from windrows

Manned or unmanned harvesters are sometimes required to pick up windrows of crops which have previously been dug and laid on the soil surface to dry. The main problem is encouraging the loose material of the windrow on to the primary web; it tends to roll ahead of the share, rather than to slide up. If there is sufficient loose soil beneath the windrow the share can be removed, so that the primary web nose runs under it; this still relies on a small build-up of material in the windrow to hold the crop against the web nose, and therefore it cannot pick up all material at the end of a windrow.

Three methods are used to aid crop flow on to the primary web.

(a) The roller share is a powered rotating, square steel bar which replaces the standard share. It rotates in the same direction as the web, and is set to run beneath the windrow. The disturbance it causes is sufficient to lift material on to the web nose.
(b) The crop can be held by a light roller running just ahead of the web nose, which presses on the windrow.

(c) A paddle wheel is driven slightly faster than ground speed, which pushes the crop on to the web nose.

6.3.3 Top pulling harvesters

These are suitable for root crops that remain firmly attached to their foliage, as well as leeks, celery and some nursery stock. After the root has been loosened by a share, the plant is caught in a gripper device which pulls it from the soil.

The most common gripper system uses a pair of endless rubber belts which press together with sufficient force to hold the plant foliage. These are inclined upwards, and run at ground speed, so that they stay still relative to the plant position, but their height above the ground increases as the machine moves forward (figure 6.5(a)).

The belt is normally a heavy section V with a widened face forming the gripper surface. This face often has a light tread pattern to improve its holding properties. The drive power is transmitted to the gripper belts by their V backing which also prevents them slipping sideways (that is, downwards) off the pulleys. The plant-gripping force is provided by a number of sprung idler pulleys which press the belts together.

For root crops the share is a single point, running to one side of the root; for leeks and celery a broader share is fitted which runs beneath the roots to sever some of the surplus material. Long fingers or rotating cone 'torpedoes' pull the top material inwards and upwards for the belts to grip. For most crops the belt height is set to grip only the unwanted part of the foliage, for example, the carrot top or leek flag leaves, to ensure that the desired part of the plant is not damaged.

The topping mechanism cuts the vegetable from its top by saw blades, or sharpened discs placed beneath the belt. The position of the topping cut, therefore, is related to the height above the ground at which the plant was gripped. This is not sufficiently accurate for carrots, where the crown heights vary in relation to the soil level. In the carrot lifter (figure 6.5(b)) the top is gripped fairly near to the soil level and, before reaching the topper, the belt grip relaxes slightly and the carrot passes between a set of guide bars which catch on its shoulders and pull it downwards by a small amount. Thus, each carrot is presented to the topper, hanging the same distance beneath the belt. Beetroot leaf stalks are not cut cleanly but torn apart in an action resembling being wrung off by hand. The beetroot topper consists of two sets of intermeshing bars which reciprocate together so that leaf stalks trapped between are broken apart.

The action of share loosening and top pulling does little to remove soil from the roots, therefore mechanisms are fitted to knock or brush the root as it passes along the belt. The exact mechanism will depend on the crop. Tap-rooted crops like carrots are knocked laterally or brushed, while leeks and celery are dragged along a bar screen which is shaking vertically.

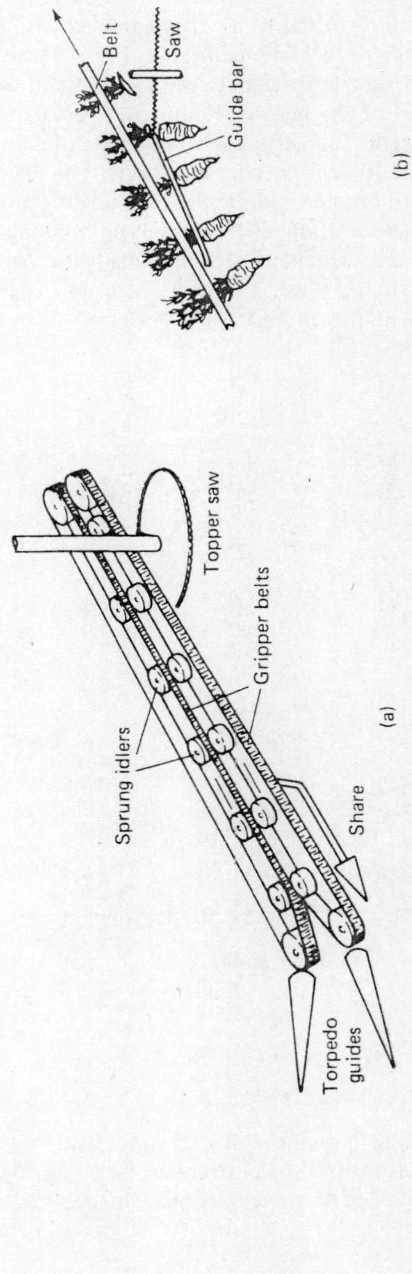

Figure 6.5 (a) Top pulling harvester. (b) Carrot top leveller.

Normally only a single crop row is taken by each belt unit, but this can include the closely spaced twin or triple rows found in, for example, early bunching carrot production.

An alternative to using twin belts is to use a large-diameter wheel with a broad edge (drum) in the place of one belt; the other belt wraps around part of the drum circumference (figure 6.6). The main advantage of this system is that the grip force between the belt and the drum edge is achieved by tension of the belt, not numerous jockey wheels, which is a simpler and cheaper construction; however, the drum diameter normally precludes multiple row configurations. The drum runs at an angle to the ground to provide a limited lifting height (about 750 mm), but nowhere near the elevation of the belt type, making topping and soil-removal mechanisms difficult to fit. The main use for this type of lifter is for nursery stock, where the drum size and angle allows sufficient elevation to a platform from which the operator can take the plants for bunching or crating.

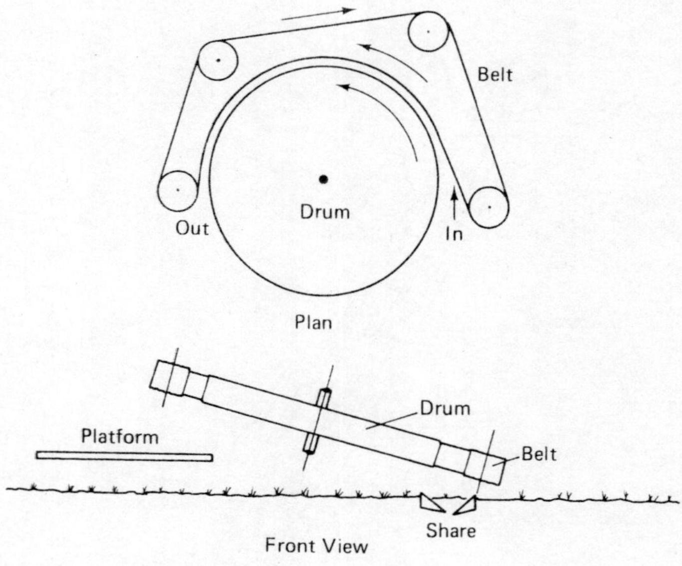

Figure 6.6 Drum and belt lifter.

6.4 HARVESTING VEGETABLES ABOVE SOIL LEVEL

The range of vegetables growing above ground is as diverse as those below, but many have their own peculiar handling and harvesting requirements, which has led to many specialist single-purpose harvesters.

6.4.1 Stem-cutting systems

6.4.1.1 Ground-level cutting

The simplest form of harvesting cabbage, brussels sprouts and cauliflower is to cut them off just above soil level, and elevate the harvested vegetables to a container. The cutting mechanism is usually a toothed disc, either arranged singly or, better, in pairs with one blade slightly higher than the other so that they can overlap to ensure a full cut. Sometimes an endless bandsaw is used, but in most conditions the build-up of mud and sap on the blade guides prevents its correct operation. The preferred cutting action is sawing, rather than chopping, with a two-bladed rotor, as this is less likely to split the stem or throw the cut head away from the elevating system. These harvesters normally elevate the crop to a container or trailer, using either gripper belts or an inclined web elevator. The conveyor handling and jumble loading in the container causes damage, therefore these systems tend to be used only for produce destined for direct processing, rather than for fresh market or storage. Jumble-loaded sprouts produce an extra task for the packhouse operators – the sorting out of stems for stripper feeding.

6.4.1.2 Cutting after lifting

To overcome problems associated with cutting low-growing cabbage and feeding it on to a conveyor, one system uses a modified gripper belt harvester where the whole plant is dug out and the root sawn off by the 'topper'. The production model uses rotating torpedoes to feed the head into the belts, which can cause significant damage. The space required for the gripper belts and torpedoes on either side of the head can require wider row spacings, leading to an uneconomically low plant population.

6.4.1.3 Selective systems

Many brassicae and lettuce crops do not reach a sufficiently high level of maturity at any one time to allow cutting of the whole crop in one pass. Selective harvesting systems have been the subject of much research, the majority directed towards methods for sensing head maturity. Cauliflower sensing is well advanced, but lettuces require non-contact methods to avoid leaf damage, which at the same time have to measure the size and density of the head, ignoring the outer leaves.

One cauliflower head sensor is a pair of lightly sprung arms, each with a soft roller at one end, which electronically measure the diameter of the head as it passes between them. Head diameter can be related to maturity, so that heads of over a given diameter are considered mature and suitable for cutting. The selective cut is done by a knife projecting from the end of a vertical shaft, which is rotated through one revolution by a hydraulic motor each time a mature head is sensed. At present this system is not suitable for U.K. cauliflower production as there is no space between the rows for the knife mechanism to bypass immature heads, and the sensing system relies on cauliflowers with an upstanding leaf configuration, working especially well on those that have been tied.

6.4.1.4 Full-width mowing

A full-width mower system is suitable for certain crops like spinach and mini-cauliflowers – a special variety of calabrese. The cutter head is built on to the front of a belt conveyor system which fills the cut crop into a container without further cleaning. The normal operating width is 2–3 m, the head mowing off all material within this width. This system is suitable only for crops in which all the plants mature at the same time, for example, mini-cauliflower or those that regenerate like spinach. The cut crop is more suitable for direct processing than for market presentation.

6.4.2 Pea and bean harvesting

Green peas and fresh beans can be completely harvested by mechanical means. The equipment is specific to the crop; one type, commonly called a viner, removes peas and broad beans from their shells, while a completely different principle is used for harvesting French bean pods.

6.4.2.1 Vining

The viner mechanism distintegrates the entire plant, with its pods, and then sorts out the peas or beans from the resulting debris. The severity of the viner threshing action is such as to break up only the waste parts of the plant; this has been aided by crop breeding, in producing brittle stems and pods.

Threshing is done within a large rotating horizontal cylinder with perforated walls (riddle drum). The incoming crop falls into the base of this drum, and is carried upwards inside the drum wall in a thin layer as it rotates. Near the top, the layer meets a small horizontal cylinder with projecting bars (stripping drum). This deflects the crop layer downwards on to a third rotating cylinder with large projecting bars (beating drum), which imparts sufficient force to break the pods open. Broken pod and liberated peas or beans fall out through the riddle drum walls while longer material and unbroken pods are carried around for further beating, until they can either escape through the riddle wall, or have worked their way along to the discharge end. The material escaping from the riddle drum falls on to the 'apron' beneath for preliminary sorting. This apron is a smooth belt conveyor, inclined and running so that it carries material upwards; the angle, speed and surface texture of the belt are such that near-spherical peas or beans roll back, but pod and haulm debris sticks and is carried out of the machine. Further debris is removed by sieving before the crop passes into a holding tank.

The angles of the riddle drum and apron are critical to performance, so the viner has to be fitted with an automatic self-levelling suspension to prevent field slopes affecting its operation.

Older viner systems picked the crop plants from a row (swath), which had been cut a short time before. The operations of cutting and vining were kept separate, as the cutter bar tended to be the least reliable part of the system, and it was more economic to have a stand-by cutter machine (swather) than to hold up the viner each time a swather failed. Modern viners operate on the 'pod stripper' principle, involving a horizontal rotor fitted with strong steel combs to pull the pods off the haulm. The rotor revolves at a speed sufficient for the pods pulled from the plant to be thrown backwards on to the intake conveyor. In addition to greater reliability, pod stripping reduces the amount of waste material passing through the viner, and allows a greater throughput of useful material.

6.4.2.2 Green beans

Harvesters for dwarf French beans also involve the stripping principle, whereby the bean pods are stripped from the standing plants by powerful revolving combs. This action also pulls off large amounts of leaf and stalk debris, which are removed by passing the stripped material through an air blast of sufficient velocity to blow the light material away from the heavier pods. Many harvesters have three stages of pneumatic separation with the crop tumbling from one conveyor to another between each stage, to help break up clusters of leaves and beans.

6.4.3 Brussels sprout harvesting

6.4.3.1 Stripping methods

Removal of sprout buttons from the stem is the most important part of harvesting this crop, four methods being worthy of mention.

(a) The stem passes through an orifice with a conical ring forming its rim. The shape of this rim has been carefully designed to form a small wedge which levers the button off the stem. To start the operation, the lower 100 mm of stem has to be cleared of buttons by hand, or by inserting it into a rotary ring saw. When the bared end is pushed into the orifice it engages a pair of gripping rollers which then pull the stalk through.

(b) A set of small knives rotates around the stem to cut off the buttons. The knives are mounted on to the rotating drive ring by a series of levers, which allow them to closely follow the stem diameter. They are held on to the stem by springs on some models, and centrifugal counterweights on others. To insert a stem, the knives are forced out to their maximum diameter by a foot pedal-operated cam. After insertion, the knives are allowed to close back to stem diameter, and the stalk is pushed against the knives until sufficient stem protrudes behind for the gripper rollers to hold. Knife pressure against the stem is critical; if too great it cuts into the stem, if too light it slices into the sprout. The centrifugal type tends to be slightly better in this respect, as spring tension can vary with stem diameter and degree of metal fatigue. Much effort has been applied to finding the optimum knife shape and attitude to make the cut without the knife being deflected inwards or outwards.

Attempts have been made to feed the stem automatically into the knife mechanism; this would both increase productivity and remove a hazardous human operation. The critical factor is to centre the stem automatically, irrespective of its shape, size or whether it has leaves remaining. Although some good prototypes have been developed, the system is not yet commercial.

(c) A variation of the rotary knife stripper uses the same knife configuration, but the knives are oscillated at high frequency rather than being rotated. Knife pressure is provided by a flexible plastic diaphragm bearing against it. As the head does not rotate, a simpler linkage is possible for opening the knife head when inserting a stalk; this is also claimed to be less hazardous to the operator.

(d) One system completely different from the foregoing is based on a similar threshing mechanism to a pea viner, whereby sprout stems are tumbled inside a drum system, causing the buttons to be knocked off. This system does not depend on each stem being manually presented to the stripper, and so is suited to the random presentation from mechanical cutting and elevating. As the buttons suffer some damage, which could develop into unsightly marks by the time they reach market, this process is used only for processing crops, where blanching soon after harvesting prevents the damage becoming apparent.

6.4.3.2 Harvester configurations

The latter system is built into a large self-propelled chassis, carrying its own cutter bar and pick-up system, that can harvest from a standing crop of sprouts with only one operator. The orifice or knife types are built up into a variety of machines.

(a) As a free-standing unit for packhouse use. The stripper head is built into a single head unit with its own drive, operator position and controls. The head is positioned so that the stem is pushed through horizontally at a comfortable working height which allows buttons to fall into a container or on to a conveyor. Damage to some buttons and leaf debris preclude the stripped product going directly into a market net without inspection and some size grading.

A stripper's ability to cope with leaves still attached to the stem depends on the design of the head infeed. The most common reason for failure is that the operator cannot see where to centre the stem in the head; this can be resolved partially with practice, and also the use of safety guards which double as guides. If a stripper can work on a stem with leaves, this obviates the need for field labour for deleafing.

(b) As a single packhouse type unit mounted on the tractor linkage. The drive to the head is derived from the tractor, but via a generator, as most machines retain the 240 V electric motors. The electric drive has the advantage of better stop/start control for those heads that require stopping for stalk insertion, and the avoidance of complex drive trains to the knife head and gripper wheels.

Taking the stripper to the field avoids the need to transport stems, and the separate labour requirements for stem loading and carting. The tractor can move the stripper close to the stems to be cut and, if fitted with extra low gearing, it can creep forward at a speed to suit cutting.

(c) For larger operations, as three or four single units mounted on a trailed chassis. The trailer can also contain conveyors for stems and cut material, trash sorters and a large-capacity container. Stem supply to the stripper operators takes many forms, the most common being either for a gang to go ahead to cut and heap the stems, the harvester driving from heap to heap, or to use a movable intake conveyor which traverses the area being cut. The conveyor can be up to 4 m long. One end is pivoted from the harvester, the other being supported by a castor wheel. The harvester is used stationary, with the cutting gang working a block 3–4 m wide beside it, and placing the stems on to the conveyor which is pulled close behind.

(d) As a more sophisticated system in which the individual operator-fed heads are placed at crop level, on an assembly that also carries a mechanical stem cutter unit. The machine moves forward sufficiently slowly for the operator to grasp each standing stem as it is severed by the cutting jaws, and feed it into the stripper head. The harvester has one cutter/stripper unit per row, but as it is wider than normal row spacing, these units have to be staggered in the multi-row harvester, as there is insufficient room to place them side by side. To allow for variations in row width, and slight steering error, each unit can pivot laterally so that its operator can move it sideways to follow his row.

The latest harvesters are based on three or four cutter/stripper units mounted in a self-propelled chassis which normally runs on crawler tracks for operation in bad soil conditions. These units do not require a separate driver, as the driving controls are placed next to one cutter/stripper unit. When working, the tracks are steered directly by signals from that unit, as the operator follows his row. The very slow forward speed allows this operator ample time to make the normal driving control actions, in addition to feeding his unit. As all the cutting and stripping is carried out within the confines and protection of the machine, it is possible to harvest the crop in weather that would be unsuitable for working out in the field.

6.4.4 Mobile packhouse and crop-collection aids

These are aids to field labour efficiency rather than mechanical harvesters, their function falling within one of two categories.

(i) Aids to the workers in the field when cutting the crop, normally based on mobile conveyors mounted on a tractor or trailer, and spanning several rows of crop. These save the cutter from having to carry, or worse, to throw, each head to a collection point. The machine travels at a speed commensurate with the cutting rate of the operatives, who work immediately in front or behind it, and place cut heads on to the conveyor, to be carried to the body of the machine for packing into a container.

In addition to improving working practice for the cutting gang, damage caused by throwing is eliminated, and the heads go directly into the container without coming into contact with soil. Crop collectors range in size from tractor-mounted units with 2–3 m of conveyor projecting either side, to large self-propelled carriers with 10 m conveyor booms. Even larger machines are used in other countries, having their conveyor supported by wheels at intervals along its length. By using a conveyor crop collector in multiple harvest crops, much of the crop area is kept free from tractor wheel travel, thus preventing damage to crop and soil structure. When using wide machines it is possible to leave unplanted trackways at intervals across the field, the crop lost in this area being compensated by the non-trafficked areas.

(ii) The mobile packhouse – an extension of the crop collector, with facilities for trimming, grading and market packing. The packhouse needs a vehicle with sufficient floor area for packing material and finished pack storage, as well as for grading, trimming and packing to take place. The requirements for these areas are as follows.

(a) Working area – where the operators sit or stand around conveyors bringing in produce and taking out waste. The normal working method is for an operator to inspect, trim, grade and pack each item, so saving on the space required for flow line operations. This area has to be illuminated at a level that is adequate for inspection and, in most cases, heated to enable working to take place in winter.

(b) Packed produce storage – normally on pallets so that rapid transfer can be made to transport. There is sometimes a need to hold two or more grades, either all on palletised market packs, or with some in bulk containers. The storage area should be sufficient for the machine to traverse a field length before unloading becomes necessary, although large fields can have transverse access roads at intervals for this purpose.

(c) Raw material storage – containers, pallets, wrappings, labels etc., and there might also need to be a separate area for erecting cartons.

These materials are normally light but bulky, and tend to be carried on outrigger racks or the roof of the main machine.

The mobile packhouse can resemble a small 'factory on wheels', with a generator supplying mains power for equipment motors and simple mechanical trimmers and sorters as well as lighting. The only operation not normally possible is washing, owing to the weight of water needed and the supply of clean water.

This system offers the producer a means of packaging a crop for market in clean and comfortable conditions, without involvement in transport of raw crop and waste. It is becoming increasingly used by specialist growers who have to find suitable land over a wide geographical area.

6.5 FRUIT HARVESTING

6.5.1 Top fruit

Apples, pears, plums and other fruit grown on trees are normally referred to as 'top fruit'. They are delicate and easily sustain sufficient damage to render them unmarketable. They are also distributed throughout all parts of the tree structure on fruiting stems which, at harvest, already carry the bud spurs for the next crop, and hence must not be damaged.

Mechanical fruit harvesting is possible because fruit becomes less strongly attached as it ripens, and can then be removed by gentle pulling or shaking. Most systems also require a special tree shape and size.

6.5.1.1 Pulling systems

These are based on a pair of inclined belts, similar to those used in top pulling harvesters, one of which travels down each side of the row of bush trees, carried under a large arched chassis which spans the trees. Long curved fingers are fitted to the belts, which are arranged so closely as to form a spaced bar conveyor. The conveyor fingers project inwards towards the tree centre and, as the machine moves forward, rise through it by virtue of the inclined belt track. The branches and fruiting spurs are drawn gently through the gaps between the fingers, but fruit is held and pulled off. The fingers are slightly curved so that detached fruit rolls along them towards the belt centre, where they lodge and are carried upwards for discharge into bins.

6.5.1.2 Shaking systems

(a) An early development used a powerful reciprocating ram which gripped the trunk to shake the whole tree. Prior to shaking, a circular sheet was placed around the tree to catch the fruit. As some fruit must fall from the topmost branches on to the sheet, this method can cause considerable damage, making it more suitable for processing fruit.

(b) The more widely used shaker system acts directly on the fruiting laterals, using combs of vibrating fingers. The combs are mounted on revolving drums or conveyors, which are freely turned by the branches as the harvester passes. Vibration is created by eccentric weight rotors running inside the drums or conveyor rollers, so that they revolve in a series of tiny jerks, rather than smoothly. The vibration energy is transmitted to the fruit laterals by the fingers. The fruit is not knocked off by the fingers tapping the branch, rather the vibration causes the fruit to swing at its natural frequency, like a pendulum, on an ever widening arc until it breaks free. The harvester either straddles the tree or works between two rows, acting on the half of each tree adjacent to it. The finger drums or conveyors are inclined outwards from a point near the trunk, and run at a height that allows the fruiting laterals to be gently caught by them. Immediately beneath run the crop-collecting elevators, which catch the fruit with the minimum drop (figure 6.7).

6.5.1.3 Cider apples

Cider apples can be allowed to ripen and fall naturally, thus require only to be picked up. Specialist picker machines are available. Some are based on large rotary brushes sweeping into a hopper, but these tend to pick up orchard debris in addition to the apples. A better system uses a gange of flexible discs which rolls across the ground. The disc spacing and stiffness are sufficient to allow the apples to become wedged between the discs. These are scraped out into a hopper carried behind the disc gang.

6.5.2 Bush fruit

Mechanical harvesting systems for blackcurrants and raspberries are also made possible because the fruits detach easily when ripe. Two methods are used to take the ripe fruit from the bush.

6.5.2.1 Combing

Early designs of blackcurrant harvesters relied on passing whole fruit-bearing stems cut from the bush through a series of rotating drums fitted with comb-like fingers. These broke off the fruiting clusters, but together with a great deal of leaf and twig debris. Light debris was partially removed by passing the stripped material through an airblast separator. These machines dictated a two-year growing system for the bushes, the stumps being allowed to regenerate in the intervening year.

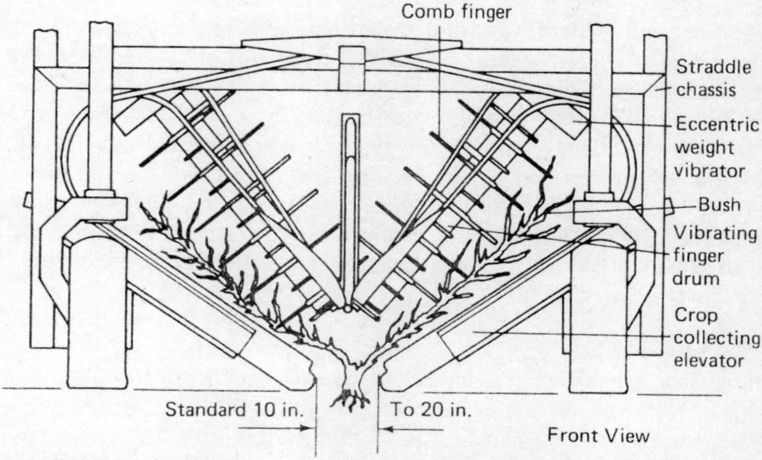

Counter-balanced elevator shoes will move round bushes and eliminate driver error

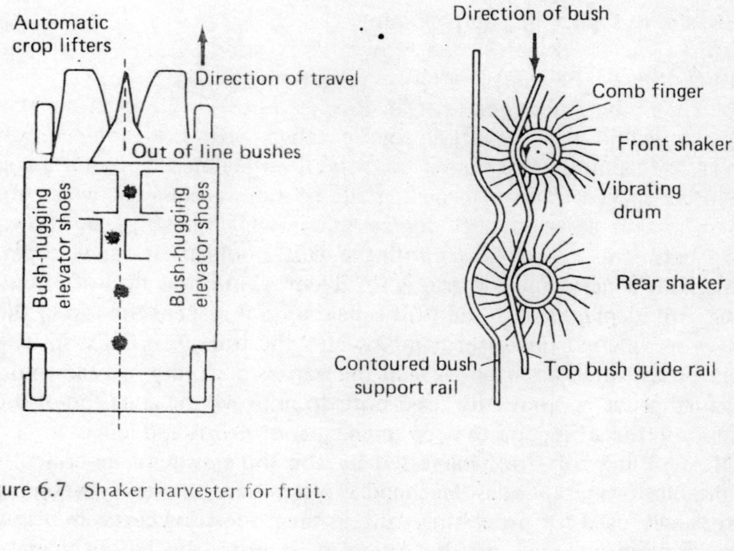

Figure 6.7 Shaker harvester for fruit.

6.5.2.2 Vibration

Later machine types use similar vibrating finger units to the apple harvester. The soft fruit vibration system was originally designed for blackcurrants, but more recently has been adapted for raspberries. Like the apple system, vibration amplitudes and frequencies have been selected to provide the best resonant 'pendulum' effect to cause the fruit to break free. For both crops a specialist harvester chassis straddles the bushes with the combing finger banks set to suit bush shape. The fruit falls on to the inclined conveyor carried at ground level. Blackcurrants are still broken off at the stem end of their stalks, and hence still tend to come of in clusters, but in most cases the 'break point' of raspberries is between the fruit and its hull. The raspberry machine can be used to work through the crop at convenient intervals, taking only ripe fruit at each pass.

6.5.3 Strawberries

Strawberry harvesters are based on cutting the fruit-carrying stems (strigs) from the plant, complete with fruit. A small reciprocating mower knife cuts the plants. Above this rotates a finger reel (rotary comb) to lift the crop for cutting and, once cut, push it over the knife on to an inclined elevator. To obtain the largest amount of fruit, the plant is cut close to the ground, which results in large quantities of leaf and stem debris being harvested also. The debris is blown away by controlled airblast separators. A platform at the rear of the harvester carries the trays which receive the fruit, and has room for an operator to assist with their filling and handling.

This is a non-selective 'once over' harvesting system, as all the fruiting parts of the plants are mowed off in one pass.

Much of the fruit coming from the harvester is still attched to its strigs. A second machine is used to effect their separation, an operation known as 'capping'. The capper consists of an inclined conveyor made of closely spaced transverse rods, about 15 mm in diameter, which are driven so that adjacent ones contra-rotate. The thin strigs are drawn down between a pair of rollers until the fruit shoulders sit against them. If the fruit is very ripe the grip is sufficient to pull the strig off, complete with its plug. If not, the fruit is held until it reaches the top of the conveyor, where a finely set bandsaw cuts the fruit from its strig. The capper uses water flotation to feed the berries on to the rod conveyor, and further water sprays are used both to improve the cohesion of the rollers and the strigs, and to keep them clear of debris and juice.

Most of the soft fruit harvested by the three systems described is unsuitable for market sales. Mechanical systems are, however, becoming increasingly used for processing fruit, as their operating costs are below that of hand picking, and the fruit can be processed before damage deterioration can occur.

6.6 NURSERY STOCK

While small seedling trees can be harvested with general-purpose vegetable undercutting machines, larger trees require special equipment. This can be either a specially built undercutter or diggers capable of extracting large trees with a viable portion of their rootball.

6.6.1 Whip and bush lifting

6.6.1.1 Undercutting

One of the major impediments to using vegetable undercutters on nursery stock much over 600 mm high is that there is insufficient clearance under the tractor for the plant to pass without damage. For taller stock an undercutter is used where the share is set to run outside the tractor wheelbase (offset). The share is wide and bowl shaped to produce a soil slice of semi-circular cross-section, to avoid cutting through too much of the main root system.

The share is carried under a large sideways projecting toolbar, which might be arched above the share for added clearance. The simplest machines have a fixed share; this can have a high draught requirement in some soils in a dry season, which can be reduced by using a vibrating share system so that part of the cutting force is provided by PTO power. In either case the offset share results in an unbalanced draught on the tractor, making it difficult to steer in a straight line. To overcome this, a tine is often fitted to the opposite end of the toolbar to produce an equalising draught force. This adds to the total load on the tractor, but this disadvantage is outweighed by the ability to follow the crop row.

6.6.1.2 Gripper belt machines

These are similar in principle to the gripper vegetable harvester, except that no topper is used and less importance is placed on root soil-removal mechanisms. The drum and belt type (figure 6.6) are more commonly used for tree seedling lifting, although the twin belt types appear to offer a higher output for larger nursery operators. The harvester either carries a platform where an operator can fill the plants into transit bins, or is fitted with a string tier which forms them into neat bundles.

6.6.2 Full trees

Mature trees, up to 250 mm girth, can be lifted with an intact rootball by specialist machinery. Most of these machines are based on a large curved blade which is pushed through the soil beneath the tree. The blade is mounted on hydraulically powered arms which move it in a circular motion to cut a spherical section of soil, containing most of the tree roots (figure 6.8).

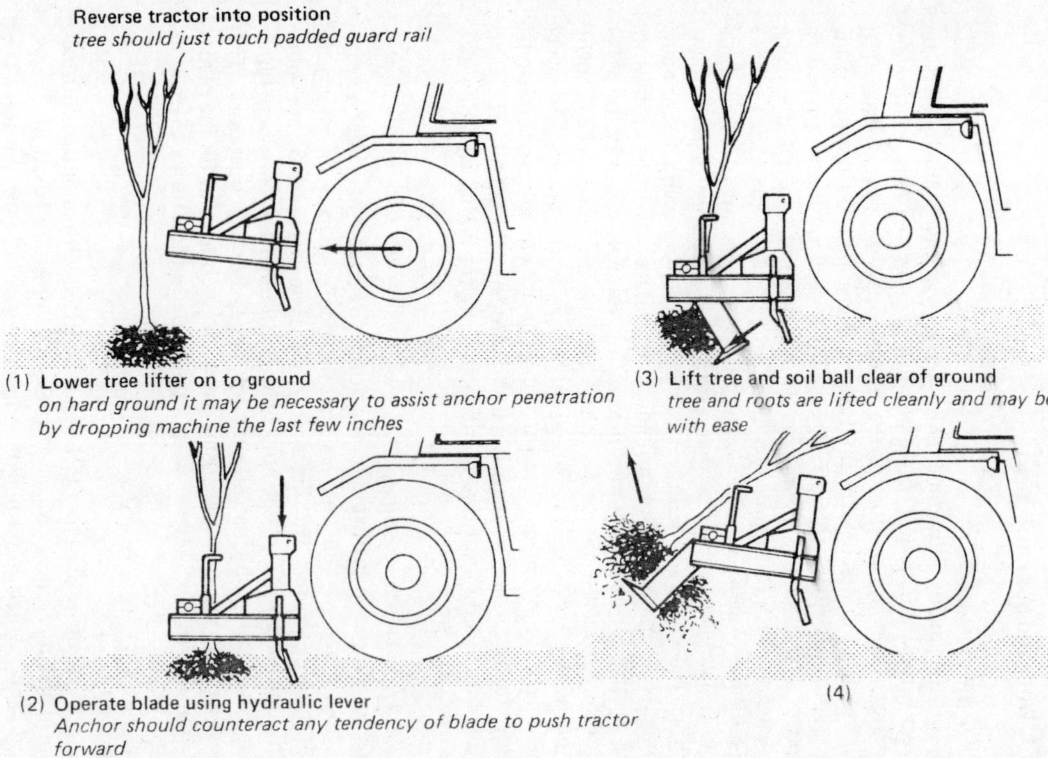

Figure 6.8 Tree lifter.

On some machines two blades are used, one each side of the tree, which close together beneath it, other types use one large blade which cuts out the rootball from one side. There is normally some means of supporting the trunk during handling; if suitable grippers are fitted the tree can be held up for sacking to be wrapped around the rootball.

Many lifters can also be used for replanting the tree by scooping out a hole of the same shape as the rootball and, where grippers are fitted, handle it into the hole also.

Larger trees can be dug out with excavating tackle and handled with hydraulic arm cranes.

6.7 CROP PHYSIOLOGY AND MECHANICAL HARVESTING

Mechanised harvesting systems cannot be considered without reference to their effects on the crop, or their complete interaction. There are many aspects of machine/crop interaction, the main ones being considered below.

6.7.1 Crop breeding

Many crops have to be bred specially for mechanical harvesting. Desirable characteristics include the following.

(a) Maturity period – which dictates whether one-pass harvesting is possible. Pea, bean and brussels sprout crops have single maturity; cauliflower and many soft fruits do not, as yet.
(b) Foliage structure – which determines how easily those parts of the plant that form the desired crop can be separated from those that do not. Peas and broad beans have haulm and pods which break easily to release their contents; sprouts have leaves which detach easily at the stem.
(c) Growth habit – which allows the required portions of the plant to be treated mechanically. Examples are: strawberries which carry their fruits on long, erect strigs; straight stemmed sprouts; fruit bushes and trees which naturally, and by pruning, form the ideal canopy shape. Many crops still lack sufficiently uniform top or root height to allow machine trimming from a soil datum; these include celery and carrots.
(d) Protection – the possession of sufficient discardable outer foliage to protect the desired inner parts from the rigours of mechanical handling. For example, cauliflowers where the leaves cover the curd, hard cabbage where soft outer leaves protect the heart.

6.7.2 Damage

There is an inherent damage capability in hand-harvesting systems, as well as in mechanical harvesting; however, the latter is often more severe and likely to have less chance of being observed and corrected. Damage exists in two forms: external, caused by cutting, gouging or abrasion; internal bruising caused by pressure or impact forces. The former is more easily detected, but more obvious in fresh market produce. Bruising often has a delayed visual effect, and consequently cannot be detected and rectified at source. Bruising might be enhanced by certain conditions; for example, potatoes develop characteristic blue/black areas under conditions of warmth and high humidity. This means that a 'perfect' ex-farm sample of washed, pre-packed potatoes can develop ugly black lesions in the environment of a supermarket.

Damage detection and warning systems are under development, but for the present the only solution lies in careful operation and maintenance of equipment.

Damage susceptibility is also tied in with cultural conditions. For example, a carrot or cabbage grown in wet conditions with its cells full of water (turgid) will split very easily. Conversely, a potato, when less turgid, will resist splitting, but will bruise much more easily; or a carrot grown fast in hard soil might split vertically when the constraint of the soil is suddenly removed at harvest.

6.7.3 Packhouse presentation

Clearing the crop rapidly from a field should not be the sole aim of harvesting. In many cases it is part of a market preparation system, and is intimately connected with the packhouse operation. In many cases rapid field operations can result in a lower packhouse rate; examples are as follows.

(a) Damage requires more rigorous inspection and hand trimming.
(b) Jumble packing requires a product to be individually inspected (like nursery stock plants) or aligned for mechanical sorting and trimming (like leeks, celery and sprout stems). Random field loading into a bin or trailer results in a large interlocked jumble pack, which requires time and effort to separate and orientate.
(c) Soiling can occur in jumble packing, where soil from the roots of one plant gets into the upper parts of others; this is especially so in leeks and celery. In plants like lettuce, the cut stem exudes sap, which can spoil the appearance of other heads that it touches.
(d) Rubbish, such as weed growth and poor plants, requires time and effort to sort it from the good produce. Excess tops, roots and soil entering the packhouse have to be disposed of. This presents logistical problems, and can spread disease-carrying pathogens if the packhouse is served from several harvesters on different sites.

6.7.4 Growing systems

Often the ideal spacial arrangement for plant growth has to be a compromise made to suit the demands of machinery. At harvest, the plants are often at their largest, and the vegetation most dense. Most harvesters operate among the crop plants, and thus require space for wheels and cutting/lifting equipment mountings to run between certain of the crop rows. Top pulling harvesters will have a limit to the number of rows that can be grabbed by a single head. Root crops like bulbs and potatoes require to be grown in single ridges to give a viable quantity of soil to be handled.

It is often more economic to suffer a slight reduction from the ultimate crop potential, in order to lower the costs or increase the speed of harvesting.

7 EQUIPMENT FOR ESTATE MAINTENANCE

A large sector of horticultural machinery is devoted to the maintenance of lawns, parkland and other amenity areas. The tasks involve mowing, hedging, ditching, forestry work and general cleaning.

7.1 GRASS MOWING

7.1.1 Cutting system

The first consideration when selecting a mower is the cutting mechanism. Modern mowers use one of two methods, impact cutting or shearing (scissoring). Each method has its advantages and disadvantages (table 7.1).

The shearing action is most often represented by the cylinder lawnmower, which cuts in the same way as scissors, one 'blade' being the edge of the cylinder bar, the other the fixed ledger plate. The cylinder bars do not run straight along it; this would produce a very uneven drive load as each blade chopped against the ledger plate. The spiral bar pattern is designed so that some part of the cylinder is always cutting against the ledger plate. To cut successfully the blades should run very close to one another but not touch.

Adjusters are provided to take up blade wear, but eventually both the cylinder bars and ledger plate lose their sharp edges, and have to be reground. A mower grinder is a specialist machine that grinds the blade edges to leave a 'cutting rake', so that only the tip cuts, thus reducing friction between the blades (figure 7.1).

The uniformity of the cropped turf is dependent on the distance that the mower travels between each blade cutting. This factor is known as 'cuts per metre' and is determined by the cylinder speed and the number of blades that it carries (figure 7.2).

Normally mowers for amenity grass and lawns have 25 cuts per metre, those for bowling greens 40 or more. The grass must poke between the cylinder bars, to be sheared against the ledger plate. Long grass cannot get between the bars sufficiently quickly to be cut, and

Table 7.1

Comparison of cutting methods

	Impact	*Shearing (scissoring)*
Cutting action	Impact requires the grass stem to stand against the blow. Cut material falls into blades and is recut	Stem is cut without being deflected. Normally each is cut once only
Finish	Ragged or torn stems; very thin material might be knocked down temporarily and stand up later	Clean cut end to each stem, weaker ones cut also
Effects of leaves and debris	Accepts debris and the recutting action breaks it into mulch	Dense mats of debris can jam the blades. Cylinder mowers might block with large amounts
Ability to pick up material	Fanning action created by some blade arrangements lifts grass into the cutter and blows mowings into a catcher	No elevation of laid grass into the blades. Cylinder mowers can propel mowings into a catcher
Effects of hard objects	Can normally deal with small stones and branches. Most blades are flexibly mounted to bounce off large objects. Blunted blades retain their ability to cut	Branches can jam the blades. Stones can raise burrs or 'nicks' and affect the smooth shearing action. Soil can blunt the blade edges
Maintenance	Replacement of cutting tips or blades relatively cheap; can be done on site without special tools or setting up	Finger bar mower blades are replaceable but normally require riveting in a workshop. Cylinders and blades require regrinding by special machinery. Needs careful adjustment to retain maximum efficiency

tends to be pushed away by an apparently solid wall of bars; this is most pronounced in a cylinder with many bars, or one that rotates fast.

7.1.2 Rough grass mowing

Most machines for rough grass are designed for heavy use, rather than for leaving a perfect finish. Most modern ones use impact-cutting methods, either flail or horizontal rotor, although some still use the finger bar.

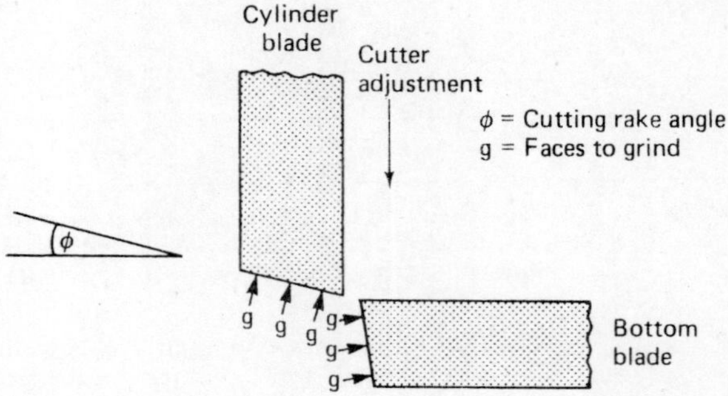

Figure 7.1 Cylinder mower sharpening.

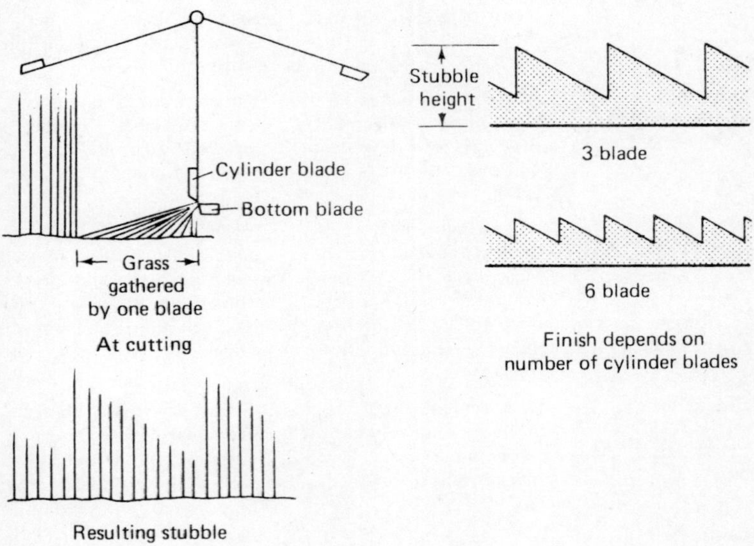

Figure 7.2 Cylinder mower cutting pattern.

7.1.2.1 Flail

Flail cutters are similar in design to the haulm pulversing toppers described in chapter 6. The main design difference is in the shape of the hood covering the flails. Instead of an outlet at the top to let the cut material escape, the hood curves down to near ground level at the rear. This traps long cut material within the hood so that it falls back into the flail bank, and is chopped many times more until it escapes at the rear as a fine 'mulch'. Mulch is preferable to long mowings as it is easily dispersed without needing to be raked up, and allows unswamped regrowth.

Flail mowers can be incorporated into small self-propelled units for small areas. For larger areas and grass verges, they are mounted on a tractor by the three point linkage, or on the end of a hydraulically operated boom. The boom mounting allows the flail head to cut areas where the tractor cannot travel, such as steep banks and ditch sides. Earlier machines used mechanical drive to the flail head but most modern ones use hydraulic drive to increase the freedom of articulation. As the power consumption is in excess of that produced within the tractor system, a separate hydraulic system with PTO-powered pump and oil cooling is fitted. Most flail mowers have a roller fitted across the rear of the hood to automatically control cutting height.

7.1.2.2 Rotary blade

The cutting mechanism is mounted at the bottom end of a vertical shaft. The cutting blades are either small replaceable mower sections fixed to a flat disc, or a crossbar with sharpened ends. On some mowers the crossbar is jointed so that the ends are free to swing back if they hit an obstruction, centrifugal force holding the bar rigid for normal cutting. The rotary blade assembly is mounted beneath a shield which prevents the operator accidentally falling into the blades, or stones hit by the blades from flying outwards.

The shield of pedestrian-controlled machines can have a tangential outlet chute to discharge cut material to one side, or a rear chute to divert the mowings into a removal box. Some machines have no specific outlet point so, like the flail machines, the grass is recut until it is sufficiently mulched to escape beneath the hood edge.

Rotary mowers can range from those with single rotors of 200 mm diameter, powered by small mains electric motors, to multi-rotor units cutting over 4 m in one pass.

Some domestic-sized mowers have specially shaped blades to generate a downwards air blast which is channelled beneath the shield to form a cushion of air on which the mower floats. Others might have blades that create upwards air currents to lift trimmings and rubbish into a collector box.

Rotary mowers for orchards have a main unit towed behind the tractor which carries independent 'wing' units to one or both sides (figure 7.3). The wing unit is mounted on a linkage which allows it to deflect on contact with a tree, so that it can cut the grass between each tree without the driver needing to manoeuvre around it.

7.1.2.3 Cutterbar

This uses a shearing action cut by a reciprocating knife fitted with small triangular blades sliding across stationary fingers. The knife unit is run through the grass at ground level and cuts it cleanly without significant recutting or mulching. This system uses less power than the impact

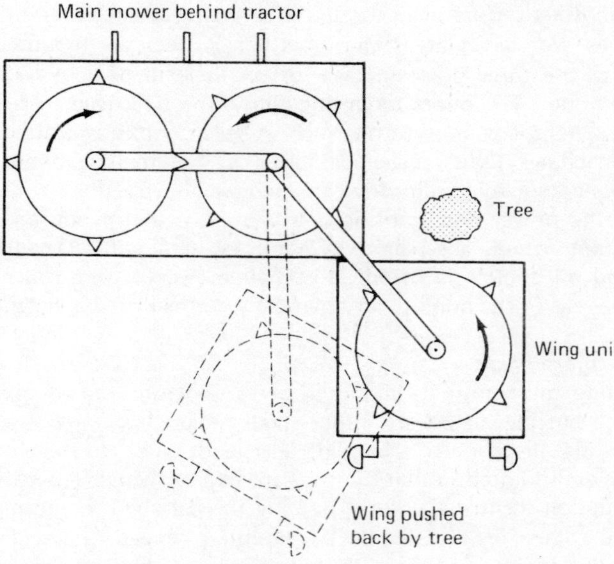

Figure 7.3 Orchard mower.

type, and is safer to use as it does not throw up stones. The knife blades have to be kept sharp, although it is common to have the reciprocating knife as a quickly replaceable item so that resharpened ones can be fitted at intervals. The system is less tolerant of debris, and thus is being superseded by the impact types.

Many cutterbar mowers are built for tractor mounting, but pedestrian-controlled versions (commonly called power scythes) are available. These are usually based on a single axle powered chassis unit, carrying a cutterbar at the front. The cutterbar height is normally controlled by small ground contact skids but, as the machine is balanced on its driving axle, the operator can easily lift it over obstacles. The power scythe is extremely manoeuvrable and can be used in areas inaccessible to four-wheeled vehicles.

It is becoming increasingly common for the power scythe to be capable of accepting a variety of cutting heads, in addition to the normal cutter-bar. These can include flail and nylon line systems.

7.1.2.4 Chain

The cutting element is a heavy chain, rotated fast enough for centrifugal force to pull it into a stiff horizontal arm. These machines are used for cutting very heavy scrub as the chain will deflect from rocks or large stumps without sustaining damage. The finish is very torn and rough, but is normally sufficient for the circumstances in which it operates.

7.1.2.5 Nylon line

These use a thin nylon line which is pulled into a 'blade' by centrifugal force, in the same way as the chain machine. The rotational speed needed to pull out the nylon line is considerable and results in a tip speed of 100 m/s which is sufficient to impact cut grass and other soft slender material. The line is, however, easily deflected by slightly harder objects, and thus is ideal for trimming grass right up to walls, kerbs or posts. Hard materials wear away the line tip which is replaced by spooling out more line. On some machines the line can be fed out while the head is in motion. Drive is by a universal type electric motor, or a small high-speed petrol engine, the latter offering greater scope for areas of operation.

7.1.3 Lawn mowing and mowing other fine grass areas

These can be maintained by using certain of the impact-cutting actions, notably flail and rotary blade, but more commonly use a shearing action cylinder cutter. In all types the cutter unit is mounted within a chassis which is able to closely control its height above the surface.

Often the chassis runs on a full-width roller at the rear, which lightly bends the grass stubble in the direction of travel. This gives rise to the much vaunted 'striped' finish, which is caused simply by more light being reflected from grass bent away from the viewer than from the cut ends pointing towards him.

Many forms of cylinder mower are made, from domestic hand-pushed machines to multiple 'gang' units, consisting of up to nine 750 mm wide mower heads, arranged in V formation. This latter method is preferable to one very wide cylinder unit, where it would be impractical to build sufficient stiffening into the cylinder or the ledger blades to prevent their centres deflecting from the prescribed gap. The gang unit is also articulated horizontally, so that each unit can closely follow ground contours. Modern gang mowers are built into a self-powered or towed chassis, which allows the gang to be folded quickly for road transport between sites.

7.2 SURFACE-COMPACTION TREATMENTS

This is normally necessary on established grass. Compaction can result from mowers and other wheeled traffic or, more often, from trampling in wet conditions. The remedy often lies in loosening the top 50–150 mm of the turf to allow surface water to percolate. This can be done in many ways, but the main criterion is to leave a smooth, firm surface, free from large divits, in contrast to crop-growing area treatments which are to be followed by recultivation. In drought conditions surface loosening can aid penetration of irrigation water.

7.2.1 Tines

The tine type 'subsoiler' can be used. It should be fitted with wide wings and run shallowly, about 250–300 mm, and must be fitted with discs to neatly cut the turf ahead of the leg. Its passage will often result in 'lips' of turf either side of the leg slot, which require pressing down, and an integral roller following each tine can be used. Operators must ensure correct power and ballasting of tractors to avoid tearing the surface through wheelslip.

7.2.2 Slitting

Slitting machines consists of shafts fitted with thin steel blades which penetrate the soil surface as they rotate. They are effective in the top 50–100 mm of turf and, if operated correctly, will not tear out divits. In hard conditions, added weight might be needed for penetration, and many machines are fitted with a tray to hold the added ballast.

7.2.3 Spiking

This involves making small holes, 75–100 mm deep, through the compacted surface. The holes are made with circular tines, either solid or hollow, mounted on a rotor. The solid tine is stronger but all the soil it displaces is forced laterally and this can compress the hole sides and prevent water moving in from the root zone. The tines are mounted on the rotor so that they move into the soil vertically.

7.3 HEDGE CUTTERS

Mechanical hedge cutters use one of three methods, scissoring, impact or sawing, each tending to be suited to a particular branch size and finish.

7.3.1 Cutting methods

7.3.1.1 Cutterbar

This involves the scissoring action of a heavy-duty version of the reciprocating grass mower. Some systems use a single blade and ledger plate but better performance is obtained with two reciprocating knife bars, each fitted with triangular blades, which slide over each other. The double reciprocating action cuts at faster forward speeds with less tendency to block. As each knife bar is driven from opposite ends of the drive crank, the out-of-balance forces are significantly reduced and thus vibration is virtually eliminated. This is an advantage in making hand-held hedge cutters easy to use. The double knife involves more moving parts than a single knife, which can make initial purchase and servicing more expensive.

The reciprocating knife is normally sufficient to cut fresh growth up to 20–30 mm in diameter, depending on machine size. In many cases the blade is able to cut any green branches that can pass between the fingers. As hedge growth is thicker than grass it does not require the same fine blade gap for shearing it, rather there is an advantage for a gap that allows sap to flow between the blades instead of binding them. Because of this, a sharp hedge trimmer will appear blunt if used on grass. Shearing action leaves the best finish to a hedge, with the minmum of roughly broken stems and leaves. Reciprocating knife trimmers are best used for taking light annual growth off regularly trimmed hedges.

7.3.1.2 Flail

Impact trimmers are very similar to flail grass mowers, and often the same unit is used for both. The flail cutter will deal with growth up to 30 mm diameter, the flexible flail mounting allowing it to bounce away from anything too hard to cut. The hood which closely envelops the flail rotor deflects trimmings back into the flails so that they are recut several times, and emerge as a fine mulch which is lost in the undergrowth. In many cases this removes the need to clear up trimmings. The flail-cut hedge has a very ragged appearance directly after cutting, with white, bark-stripped stems very prominent, but this effect is improved by leaf regrowth. The impact action does not have any long-term effect on hedge growth. The flail is suitable for hedges with several years' growth, and can trim these severely if necessary.

7.3.1.3 Saw

Saw cutters use large-diameter circular saw blades running against the hedge. The blade is operated at normal wood-sawing speed (peripheral speed 20–30 m/s), and the resulting cutting action is a combination of impact on leaves and small material, and sawing on larger branches. Saw tooth size and spacing vary with wood thickness, from slasher blades with a few large teeth for light growth, to a large number of small teeth able to deal with quite large boughs. The saw is thus suitable for general trimming, cutting back hedges with many years' growth, or completely demolishing them. The blade teeth have to be sharp for cutting; many are tipped with wear-resistant material, such as Stellite or tungsten carbide, to increase periods between resharpening. A saw-trimmed hedge initially appears slightly ragged, mainly owing to the small wood being impact-cut, but soon recovers. The trimmings are not significantly mulched and often have to be cleared up. The saw is more easily damaged by metal or concrete posts in the hedge than the flail. It is also costly to sharpen and replace. This has to be done on the blade as a whole, rather than on individual flails in a rotor.

7.3.2 Hedge trimmer configuration

Small reciprocating blade trimmers can be hand held, either driven directly from a small petrol or mains electric motor, or driven remotely from a stationary engine by hydraulic power, Bowden cable or generated electricity. The reciprocating knives on these trimmers usually have blades on both sides, so that cutting is possible by sweeping them across the hedge face in either direction. Saws and flails are too unwieldy for hand operation, although a small petrol-driven saw-blade machine is made for brushwood cutting at ground level.

Saw, flail and large reciprocating knife trimmers are tractor mounted on the end of a multi-jointed arm. The geometry of the arm is controlled hydraulically to enable the cutter head to top a hedge up to 4 m high, trim its nearside face at a distance of up to 3 m from the tractor, and even reach over to trim the upper part of the opposite face (figure 7.4).

The cutter head is hydraulically driven but, like the flail grass mower, requires a separate PTO-driven oil supply. The trimmer hydraulic system requires a large quantity of oil, up to 140 litres, to operate

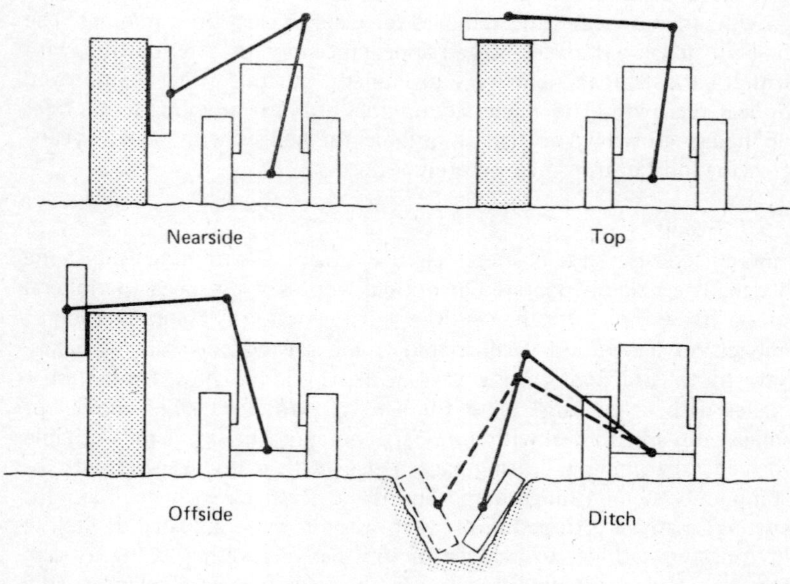

Figure 7.4 Flail mower head positions.

effectively. This is held in a large tank, mounted on the opposite side of the tractor to the arm, and helps to balance it.

The head position is easily adjusted from the tractor cab by spool valves controlling the ram at each pivot in the arm.

7.4 TIMBER CUTTING

Some of the heavier equipment used for rough grass cutting and hedge trimming can be used for clearance of scrub and even small saplings in woodland work. Work on mature trees, however, requires different, specialist tools.

7.4.1 Chainsaws

These are the general-purpose tool for felling, lopping and cutting up mature trees. The chainsaw is based upon a loop of steel chain carrying cutter teeth (figure 7.5).

The chain runs around an elongated former (guide bar), with the engine and drive sprocket at one end. In some models the chain runs directly on the curved end of the guide bar (plain nose type), while in others it is carried on an idler sprocket. As the nose end of the cutter chain is unenclosed, the full length of the bar, including its tip, can be used to cut. The maximum thickness of trunk that can be sawn is twice the bar length, as it can be cut from both sides. This compares well with a circular saw blade which has a depth of cut around one-third of its diameter, or a hand saw which has to be longer than the trunk thickness to allow it to reciprocate.

The chainsaw is normally driven by a small petrol engine, although electric and hydraulic models are available. Petrol engine machines normally incorporate a centrifugal clutch, so that the blade operates only when the engine is running at speed, and is disengaged while the saw is being started or is idling. The accelerator trigger thus doubles

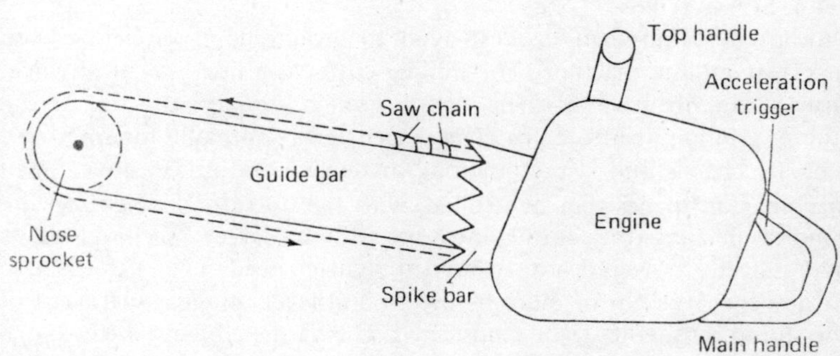

Figure 7.5 Chainsaw (guards omitted for clarity).

as a clutch to allow one-hand control. The chain rotation is such that the cutting resistance pulls the saw forward. The row of fingers (spike bar) projecting from just behind the engine hold the saw against the forward drag, and also act as pivots, so that raising the engine end of the saw pushes the blade into the wood.

7.4.2 Saw benches

These are circular saws mounted vertically in a heavy frame. The blade is driven by either the tractor PTO or an independent engine located in the frame. The common sizes for general forestry use a 600–900 mm diameter blade, and a maximum cutting depth of 250–350 mm. Saw-benches are used for timber that can be manhandled on to the bench frame, and offer a more rapid method for cutting up small timber after felling than the chainsaw, which requires each piece to be propped and wedged for safety. The sawbench can be a dangerous tool when it is being used to manually feed timber into the blade. A heavy guard should be fitted over the blade to prevent the operator falling on to it, and a 'riving knife', which is a thin fixed blade, behind the saw to prevent the timber cut from closing and jamming it. Additionally, many saws are fitted with a feed carriage which propels the timber to the saw.

7.4.3 Power pruner

For light-duty cutting in woodland and orchards, power-operated versions of hand pruners have been made. These normally operate penumatically from a small portable compressor, the air ram simply replacing the hand-force action. In this instance, pneumatic power is preferable to hydraulic, as the ram is lighter and requires only one feed pipe which is smaller and more flexible than hydraulic pipe.

7.4.4 Stump removal

Unwanted stumps can be destroyed to ground level or below by a specialist milling machine. The milling cutter is a heavy vertical wheel with a rim of sharp hardened teeth, which runs on the top of the stump, grinding it into chips. The cutter unit is normally mounted in a separate chassis unit also containing the engine and power drive. Small machines are pedestrian controlled, with the operator's weight on the handles pushing the cutter into the stump; larger machines use a hydraulically powered arm to hold the cutter head. The cutter is able to descend 100 mm or more below ground level, to clear sufficient of the stump for the site to be landscaped. The hardened teeth are essential to counteract the abrasion of soil and stones held among the roots, when the cutter is working below ground level.

7.4.5 Log splitters

If felled timber is to be exploited for firewood, larger logs will have to be split into manageable sizes. Powered splitters can be used, working either by forcing a wedge into the wood or screwing a threaded cone into it. The wedge is normally forced in under hydraulic power, hand machines using simple jack rams and larger models using tractor oil power. The threaded cone type is usually tractor PTO driven via a reduction gearbox. Both types require the logs to be cross-cut into short lengths before splitting.

Saws and other forestry cutters are potentially dangerous and must be operated with extreme care. The risks extend beyond the hazards from the cutting mechanism to such things as flying debris and falling timber. In the U.K., training in their correct use can be arranged through the Agriculture and Horticulture Training Board, and information on safety can be obtained from the Health and Safety Executive (see appendix B).

7.5 DRAINAGE

Removal of excess soil water often involves two operations, laying drains in the field area and digging ditches to convey away the water that they remove.

7.5.1 Pipe drainage

The pipe has to be placed between 0.5 m and 3 m below the soil surface. It must be laid with a slight continual fall towards the ditch outlet, irrespective of the contours of the field above. The pipes might be of clay or perforated plastic.

Laying clay and large-bore plastic pipes needs an open trench, while small-diameter plastic pipes can be laid by the 'trenchless' method, which involves a large mole plough opening a slot of sufficient width for the pipe to lie in. The pipe is fed down the back of the leg, and the slot closes again after the plough has passed. Sometimes a second applicator is attached to the leg to fill the lower part of the slot with porous material to aid water passage to the pipe. A trenchless plough can also be used to lay irrigation pipe or under-soil heating systems.

The trench method can be carried out with a 'backhoe' excavator, but this is slow compared with a specialist drainage machine, which uses endless conveyor, or a wheel fitted with digging buckets, capable of digging to 3 m or more. The trench is normally only 150 mm wide, unless larger drains are to be laid. A pair of vertical plates extends for a short distance behind the digger to support the trench walls; within these plates is the feeder system for sliding the pipe into the trench base, and optional porous backfilling. To save on weight, the porous backfill is supplied from a tractor-drawn hopper fitted with a conveyor-belt discharge, driven slowly alongside. The porous fill is sometimes applied directly from the hopper trailer in a separate operation, but this is possible only if the trench sides are stable.

Excavated soil is returned by a tractor fitted with a dozer blade. In light, dry soil the tractor can have an angled blade on the front, so that it can run alongside the trench, but in heavy, wet soil a straight blade is used with the tractor approaching the trench at right angles.

An important part of any drain layer is the depth control, which provides the necessary fall, free from undulations, because it is not sufficiently accurate to take the trench base datum from the field surface. Early systems used a set of 'T'-shaped sighting poles previously set out with a surveying level, so that the driver could line the top of a sight on the drainer with the tops of the 'T' poles. Modern systems employ automatic electronic levelling, using a horizontal light beam emitted by a laser device placed anywhere in the field in sight of the drainer. A receiver on the drainer picks up the laser beam, and 'sights' back along it to the instrument datum.

The trench system can lay to greater depths than can the trenchless, and also enables a grid of mains and laterals to be laid with junctions in the field, rather than having to start each run at a ditch or large excavated hole. The drainage machine destroys 1–2 m of crop along the trench line, and thus is not suitable for amenity grass areas in regular use. The trenchless system does not cause so much damage to the surface; turf can be reinstated relatively easily provided that the traction unit has not caused rutting as a result of wheelslip.

7.5.2 Ditching

Ditch digging and major reclamation operations involve heavy backhoe or dragline excavators. They can also be used for normal maintenance, although this task can be performed equally well by lighter machines. A heavy, narrow bucket is used for digging, whereas a light, wide bucket can be used for normal maintenance, which requires only the removal of accumulated mud, and can be tackled by scraping rather than digging. The excavator is normally sited at right angles to the ditch, so that it can draw its bucket down the opposite side, across the bottom and up the nearside. The bucket need only be shallow, as not much material is handled, and is wide to allow rapid progress. Holes in the bucket allow any water caught to escape. Other types of ditch cleaner have been made from time to time, some using large digger wheels running with, or across, the ditch, but most have been superseded by the hydraulic arm excavator.

The other operation in ditch maintenance is weed removal. This can be done with special mowers, the commonest being a hydraulically powered, reciprocating knife fitted to the front edge of the wide ditch bucket which is drawn across the ditch as described above, this time only cutting weed growth. The cuttings fall back into the bucket for lifting clear of the ditch, rather than back into the water. The arm-mounted flail mower/hedge trimmer can also be used to trim ditch sides.

7.6 CLEANING MACHINERY

There is often a need to clear amenity areas of leaves, mowings and other debris, and clean mud from roadways.

7.6.1 Road sweeping

The road-sweeping machine uses large rotary brushes with widths from 500 mm to 3 m. The smaller brush unit is suitable for fitting to pedestrian-controlled power units. like the garden rotary cultivator. The larger types are fitted to tractors or the larger 'ride on' groundsman machines.

The brush rotates against the direction of travel, sweeping the dirt forward. The sweeper can be angled to one side or both sides, so that the swept dirt is worked to the side of the roadway. The rotational speed is set to fling dirt clear of the bristles by centrifugal force, but not so fast that it creates a dust nuisance.

Some machines have two brush speeds, low for dry conditions, and high for wet clinging rubbish. The bristles can be of different thicknesses and stiffnesses so, for example, a soft brush can be selected for lightly sweeping a pitch, and a hard one for scrubbing mud off a road surface.

7.6.2 Leaf sweepers

These are rotating brush machines where the brush is driven in the same direction as, and slightly faster than, forward travel. This propels mowings and leaves into a shallow hopper carried behind. The brush is not normally an entire cylinder of bristles, but rather a series of longitudinal bars of bristle forming paddles, which hold the debris and flick it into the hopper. The power needed to drive the leaf sweeper is much less than for the 'roadsweeper' as it is brushing with the forward motion; thus small machines can be hand pushed, driving the brush from the ground wheels.

7.6.3 Vacuum cleaners

Like the domestic vacuum cleaner, these pick up debris by means of an inrushing air jet. The size and tenacity of material that a cleaner can ingest is governed by the air velocity it generates around the nozzle (capture velocity), not its suction pressure. Air entering a nozzle comes

Table 7.2

Capture velocity created by a cleaner with a 100 mm diameter nozzle with an air volume of 0.75 m³/s

Distance from nozzle end (mm)	*Capture velocity (m/s)*
Into nozzle	95
10	85
50	23
100	7
150	3
200	2
250	1

from a wide area around it, and the capture velocity decreases rapidly with the distance from it (table 7.2).

Table 7.3 shows the air velocity needed to convey certain materials.

Table 7.3

Air velocity for conveying material

Material	*Velocity (m/s)*
Dry sawdust	10
Limestone dust	13
Green sawdust	15
Metal from shotblasting	18
Cereal grain	19
Pulverised coal	20
Dry wood chips	20
Green wood chips	25

Materials like leaves can be equated with wood chips at 20–25 m/s, and soil dusts with materials like limestone dust and shotblasting metal at 13–18 m/s. These figures establish the air velocity in the vicinity of the inlet needed to make the cleaner effective. The cleaner in table 7.2 could pick up most materials within 50 mm of the end of the nozzle, and will pick up light (limestone) dust within 70 mm. The nozzle is normally connected to the machine by piping, which has the same or a smaller cross-section, so the air velocity in the pipework can be higher than the 'into nozzle' figure in table 7.2. If, for example, the 100 mm nozzle is connected to 75 mm pipe, the velocity in this pipe will be 170 m/s.

Velocities of this magnitude create very high frictional resistances between the air and the pipe walls. The total of the resistances due to the air entering the nozzle and travelling through the pipework results in an air pressure drop at the fan, the 'suction' pressure that the cleaner fan has to produce. Thus the 'suction' of a cleaner is not a measure of its picking-up capability, but a consequence of its design!

Material travelling along the pipe can be extracted in the catching hopper in a number of ways.

(a) A large sealed container can be placed before the fan, lined with air-filtration material to stop particles, but allowing air to continue to the fan. This method has a big disadvantage when picking up very fine material, as the filter quickly becomes blocked.
(b) The sealed container can be used without the filter lining as a first stage, to remove much of the air-borne debris, before the filter chamber. As the air enters the container, it expands to fill the greater volume and its velocity falls, as does its ability to carry material. The chamber normally contains baffles to prevent stray particles being blown straight across to the outlet.
(c) A refinement to (b) is the cyclone, a vertical cone with the air pipe entering tangentially to the top of the wall. The heavy particles spiral downwards close to the wall, while the air stays near the centre and leaves by an outlet in the centre of the top.
(d) The foregoing types clean the air before it passes through the fan. It is possible to convey the debris through the fan to a filter box or cyclone on the discharge side, but this requires a specially built unit with heavy blades to accept abrasion, and large clearances to prevent blockage.

Pneumatic conveying is very expensive on power, and even small hand-pushed units require over 2 kW to drive the fan. Inlet arrangements are either a fishtailed hood mounted on the machine chassis, to suck up debris as it is pushed along, or a large-diameter flexible 'trunk' which can be used to remove leaves from herbaceous borders without damaging the plants.

7.6.4 Blowers

These, used to 'push' leaf and mowing debris off lawns and borders by means of an airblast, are simply an engine-driven fan, blowing directly or through a flexible 'trunk'. The air velocities required in the trunk are lower than for a suction unit. As a discharging jet does not spread to the same degree as an inward jet, its effective velocity is retained for a metre or two beyond the orifice, which allows it to convey leaves without needing the nozzle to be held close to them. It is desirable that the blower has a properly shaped nozzle and is run at the correct velocity to prevent the airblast scattering the material, instead of pushing it in an orderly fashion. As the material never travels through the pipe or fan, there is no need for settling chambers, filters or special fans.

7.7 FENCING

There are two methods for inserting tree stakes and fence posts mechanically.

7.7.1 Post hole boring

This uses a large Archimedian screw auger, which is power driven by a tractor or a portable power unit.

The tractor-mounted post hole borer is carried on an extension frame from the three point linkage, and driven by the PTO through a reduction gearbox. The frame length and its mounting geometry provide sufficient lift to raise and lower the auger, while trunnions between the auger and frame allow it to remain vertical irrespective of the frame angle.

The portable unit consists of a small engine and vertical gearbox mounted on a carrying frame; sometimes this is an adaption of a chainsaw drive unit.

The auger point is made as a replaceable item, both to allow renewal when worn and to enable the most suitable point to be fitted for the soil conditions.

7.7.2 Post driving

Most powered drivers are tractor mounted and work by the falling weight 'pile driver' principle. PTO or hydraulic power is used to raise the weight and it then falls against the post top. The weight operates on a vertical guide which also carries a clamp to hold the post vertically during driving.

An alternative design uses the tractor's hydraulic power to force the post into the ground. In this type the driving force is limited to the weight of the rear of the tractor.

8 HANDLING EQUIPMENT

Every sector of the industry is involved in the movement of materials. Most of this is now done mechanically, either to alleviate the tedium of repetitive carrying, or to cope with loads greater than the human frame can handle.

8.1 TRAILERS AND TRUCKS

This sector includes trucks, barrows and trailers, all based on running gear (wheels) supporting a load-carrying platform.

8.1.1 Running gear

In addition to wheels, this includes the axle, brakes, suspension and tyres.

8.1.1.1 Tyres

One of the most important components is the tyre; it has to support the load and provide a degree of cushioning between the road surface and the axle. The load-carrying potential of a pneumatic tyre is governed by a combination of the air pressure and its carcass stiffness or 'ply rating'. An example of this is shown in table 8.1.

Road-going tyre specifications also include travel speed because the flexing of a tyre wall at speed creates heat, and if this becomes too great the tyre is damaged.

For field work the important consideration is the ability to run on soft soil without excessive sinkage, which otherwise causes high resistance to pushing or pulling. This is normally termed 'flotation'. Recent work at the N.I.A.E. has shown that, while trailer tyres are specified correctly for 'on the road' load carrying, their flotation on soft soil is poor at the trailer's rated load. Considerable benefits in reducing draught and soil damage have been demonstrated using over-sized tyres at relatively low pressure.

Table 8.1

Tyre load limits (kg) at various cold inflation pressures for trailer tyres

Tyre size	Inflation pressure Bar / kPa	1.0 / 100	1.25 / 125	1.5 / 150	1.75 / 175	2.0 / 200	2.25 / 225	2.5 / 250	2.75 / 275	3.0 / 300	3.25 / 325	3.5 / 350	3.75 / 375	4.0 / 400	4.25 / 425	4.5 / 450	4.75 / 475
3.00–16		45	48														
5.50–16		310	350	385	425	460	500 (4)	540	580	620	660 (6)						
6.00–16		380	450	475	525	570 (4)	610	645	685 (6)	720	755	790	820	855 (8)			
6.50–16		425	480	535	585	640 (4)	685	730	775 (6)	810	850	890	925 (8)				
7.50–10			470	510 (4)	550	585	625	660 (6)	710	760	805 (8)	855	905	955 (10)	995	1035	1075 (12)
7.50–16			635	700 (4)	765	825	890 (6)	940	995	1045	1100 (8)	1145	1195	1240 (10)	1295	1355	1410 (12)

Note: The numbers encircled denote the ply rating for which the accompanying load and inflation is a maximum.

8.1.1.2 Brakes

The other important component of the running gear is the brakes, both for safety and complicity with road traffic laws. At the time of writing the minimum requirement in EEC countries for tractor-drawn trailers is a brake that can be operated from the tractor seat. Modern tractor cab design has prevented the use of drawbar-mounted grab levers, and forced many manufacturers to find a means of operating the trailer brake from within the cab. The simplest system involves the hydraulic oil supply used for trailer tipping. A valve is fitted to switch the oil supply to the brake ram before running on the road, which allows the driver to apply the brakes by operating the 'tipping' lever in the cab. Newer tractors can be fitted with a brake pedal-operated valve to activate the 'tipper' oil supply.

The main drawback to this method is that it requires the driver to pre-select the braking option. This can be avoided by fitting independently operated brakes, controlled from the tractor brake pedal. Two systems exist, 'single line' or 'triple line'.

(a) Single line uses one pipe to transmit hydraulic or air pressure to the trailer brake ram. The hydraulic system is separate from the tractor's other services, and requires modification within the tractor hydraulic pack. This system has a big disadvantage if the pipe leaks or is cut, as the trailer's brakes become inoperative; that is, it does not 'fail safe'.

(b) Triple line includes a fail safe system. It runs only by air pressure: one pipe operates the brake ram in the same way as single line; the second continually supplies air at full pressure to charge a storage cylinder on the trailer; the third is connected to a valve between the storage cylinder and the brakes, and in normal use it also receives full air pressure, which holds the valve closed. If this pipe leaks or is severed and loses pressure, the valve opens and allows the full air pressure in the storage cylinder to apply the brakes.

Transport regulations in the EEC demand that heavy goods vehicles (trucks) have a fully operational secondary braking system, in addition to the parking brake, operated by a two or three line mechanism. These regulations also specify the minimum deceleration rates for each system. At the time of writing, these regulations do not encompass agricultural (horticultural) vehicles; despite this, some trailer manufacturers already offer systems that would comply. Most U.K. tractors do not have a compressor system to operate air brakes, but this can be overcome by fitting an 'air brake pack', consisting of a PTO-driven compressor, air tank and the necessary valves.

8.1.2 Load-carrying platforms

8.1.2.1 Bulk transport bodies

These take many and various forms, the most common being a flat platform with hinged sides to allow it to carry either large objects or bulk material. In many cases the platform can be tipped by a hydraulic ram arrangement fitted beneath it. The platform is most commonly hinged to tip rearwards, but some manufacturers supply three-way tipping trailers. These have a vertical ram placed centrally beneath a platform which is mounted on the chassis by special V-shaped brackets with removable pins across the top. Removing pins from the appropriate brackets frees a particular side for lifting, the platform pivoting on the opposite side. Three-way tipping is useful when working in confined areas, where the trailer cannot be reversed to its dumping position.

A trailer used to carry only loose bulk materials will not need hinged sides, and can be built on the 'monocoque' principle. In this a series of 'U'-shaped metal ribs form supports for both the floor and the walls. The sheeting is welded to these frames, and ideally contributes to the trailer's structural strength. The rear end can have a hinged tailboard, or to speed the tipping operation, an upward sloping end section or 'scow end'.

Trailer sideboards can be of different shapes and heights to suit the load being carried. Where the trailer is being used to catch the discharge from a harvester it is common to have a low side adjacent to it, and a high side opposite to prevent material over-shooting.

8.1.2.2 Specialist bodies

Many trailers are built for a specific purpose, open racks for example, to carry irrigation pipes, or a low flat platform on to which other machines can be driven.

Machine transport trailers need facilities for the machine to run on from ground level. This can be a set of movable ramps, but trailers are built that either tilt until the rear end touches the ground, or have wheel equipment on an hydraulic linkage, so that the whole body can be lowered to the ground.

8.2 CONTAINER HANDLING

Many products are now handled in large bulk lots, weighing 500 kg or more, in large boxes or bags of granular material, or as smaller packs assembled on a pallet.

8.2.1 Box and pallet specification

The engineer must be aware of three aspects of box or pallet design.

8.2.1.1 Load capacity

Pallets are normally loaded with $\frac{1}{2}$ tonne or 1 tonne of product and, exceptionally, $1\frac{1}{2}$ tonnes, as in the case of bagged fertiliser. This will determine the specification of the handling equipment. Loose-filled boxes cause the greatest problem, as materials of differing density will be carried. For example, a box 1.8 m x 1.2 m x 1 m deep, of 2 m^3 internal volume, can hold $2\frac{1}{2}$ tonnes of sand, 850 kg of daffodil bulbs, or around 150 kg of bare root nursery stock plants. One often sees a box labelled as '1 tonne' or 'half tonne'; this is normally based on potatoes, having a bulk density of 1.6 m^3/t.

8.2.1.2 Dimensions

Series of standardised pallet and box dimensions have been suggested by the bodies representing British and International Standards. These are important for pallets or boxes that circulate off the holding when, for example, sending produce to market, or carrying purchased raw materials. The standard dimensions of the pallet shown in figure 8.1 are: 1 x 1.2 m; 1.2 x 1.2 m; 1.2 x 1.8 m. Pallets to be carried on road transport should be 1.2 m long, as the maximum load width allowed on a lorry is 2.4 m, suiting two such 'standard' pallet lengths.

Pallets and bins used within any particular business can be sized to suit requirements. An unofficial standard has evolved within the potato industry of 0.9 x 1.5 m which is convenient for two purposes: four pallets fit conveniently on a farm trailer bed of 1.8 x 3 m, and one pallet holds four standard chitting trays.

8.2.1.3 Construction

One often sees a pallet or box base referred to as having 'two way' or 'four way' entry. This refers to the number of faces that the fork truck can enter (figure 8.1).

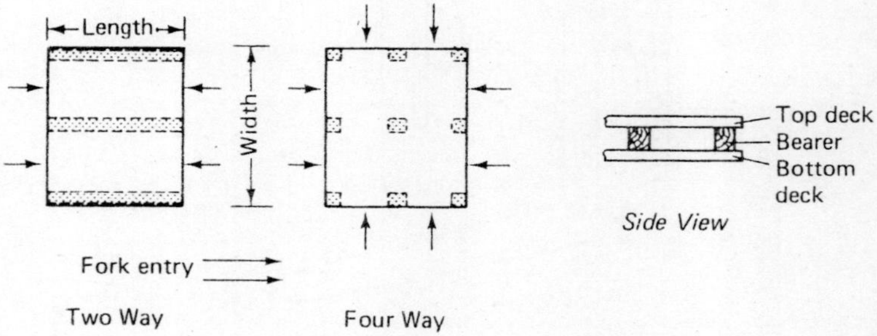

Figure 8.1 Pallet details.

The four way pallet is less strong than the two way, as there is less timber to support the top deck. Its main advantage is in improved access; for example, a four way pallet can be loaded on a lorry from the side and unloaded by a forklift running on from a rear dock. Another important factor of pallet base design is whether it is 'open' or 'closed', which indicates whether or not a bottom deck is fitted. This can affect the choice of tipping equipment, and some small forklift and stillage trucks.

8.2.2 Forklifts and other pallet handlers

These have become very widely used in recent years; they can be divided into two categories, forklifts and stillage trucks.

8.2.2.1 Forklift construction

Forklifts, as their name implies, lift the pallet on a pair of tines (forks).

(a) The forks are about 100 mm wide and taper from 25 mm thick at the rear, to about 10 mm at the front (figure 8.2). The thin tip increases the ability to slide into the pallet forkspace; the thickened rear end better resists the bending forces on the tine. Tine length can vary, but it is normally at least 1 m. A tine that is too long protrudes through the pallet that it is picking up, and can catch the next one. If too short, some of the pallet slats are unsupported by it, which can cause weak pallets

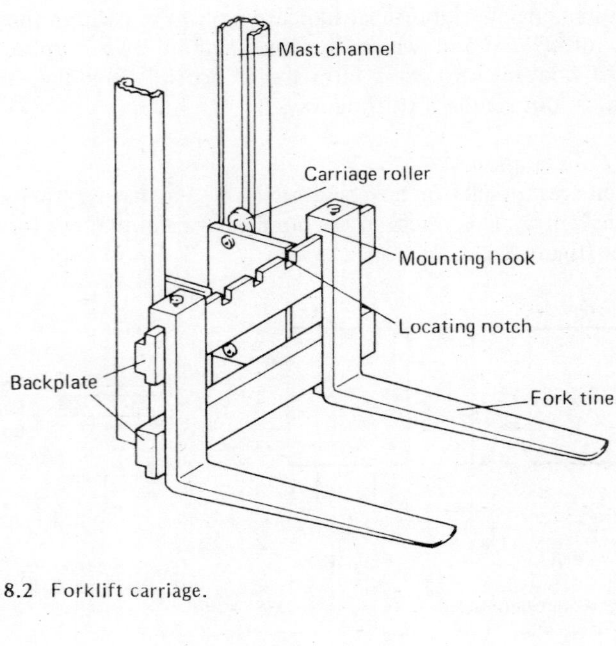

Figure 8.2 Forklift carriage.

to break. Those that are too long can have a baulk of timber laid across the rear end to prevent their entering too far; short ones can be sleeved with hollow tube. These measures can be used only if the forktruck capacity allows, as both will place the pallet weight further forward.

(b) The fork tines are mounted on a 'backplate' (figure 8.2), by a hook at the top of the vertical leg to allow tine spacing to be adjusted for pallet width, or be easily exchanged for other attachments.

(c) The backplate is mounted on either a carriage which slides in a vertical mast (figure 8.2) or the end of a hydraulically operated arm or boom (see figure 8.4 and (f) below). In all but the cheapest machines the carriage runs inside the mast channel on small rollers to reduce friction. The mast is usually made of two or more sections which telescope (figure 8.3) so that its height is minimised when the fork carriage is at the bottom of the mast, for access under low structures. The standard mast has two sections and lifts the load to 3–3.3 m; special forklifts having three or four section masts can lift to heights above 6 m. The height of the mast when fully telescoped is termed 'closed

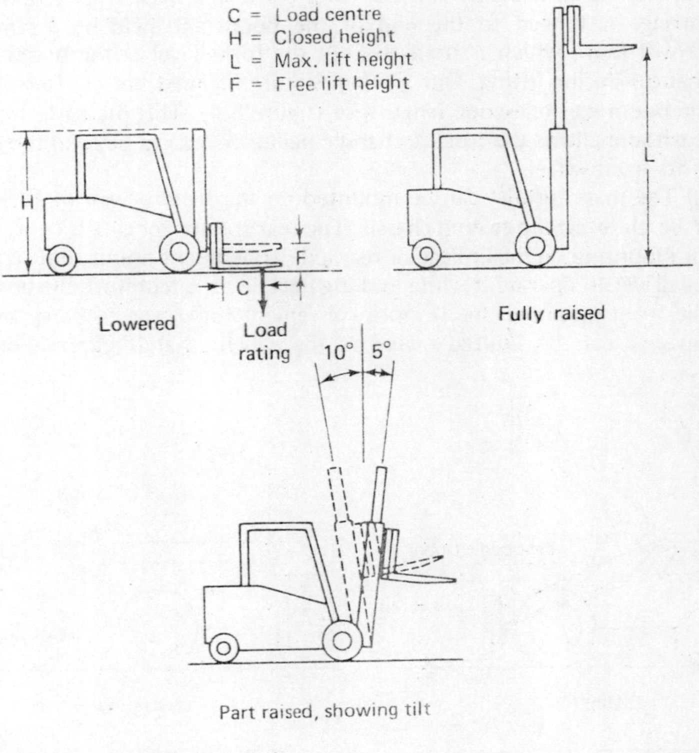

Figure 8.3 Two-stage mast forklift.

height'. The carriage lift mechanism normally allows the forks to be raised by 200–250 mm before the mast begins to extend, to allow the machine to carry a pallet within the restricted headroom. This is termed 'free lift' (figure 8.3).

(d) The carriage lift is normally hydraulically powered. To keep the ram arrangement simple its stroke length is increased by a system of chains and pulleys.

(e) Two further controls on fork position are provided by 'tilt' and 'sideshift'. Tilting the mast enables the angle of the forks to be changed slightly to keep them horizontal under varying load and ground surface conditions. Sideshift provides lateral movement of the fork carriage, so that the pallet can be placed accurately without the forktruck needing to be steered precisely.

(f) The boom lifter is a relatively modern innovation which has evolved from a loading shovel. This machine does not require the sliding mast sections and lifting chain system of the conventional forktruck, and is therefore mechanically simpler. The boom provides the lifting power to the fork head and can lift a load to the height of the base machine without any upward protrusion due to a telescopic mast (figure 8.4). The carriage is hinged to the end of the boom and held by a ram, the 'crowd ram', which is used to keep the forks level as the boom angle changes during lifting. One further feature of most boom lifters is that the boom can telescope lengthwise (figure 8.4). This increases forward reach and allows the truck to handle pallets 3–3.25 m beyond the travel of its front wheels.

(g) The mast forklift can be mounted on the front or rear of a tractor, or be an integral powered chassis. The rear tractor forklift is convenient for mounting on the linkage or rear axle attachment points, but requires the driver to operate it while looking back in an uncomfortable posture. The front mounted one is more convenient for driver visibility, but its capacity can be limited owing to the weight that it imposes on the

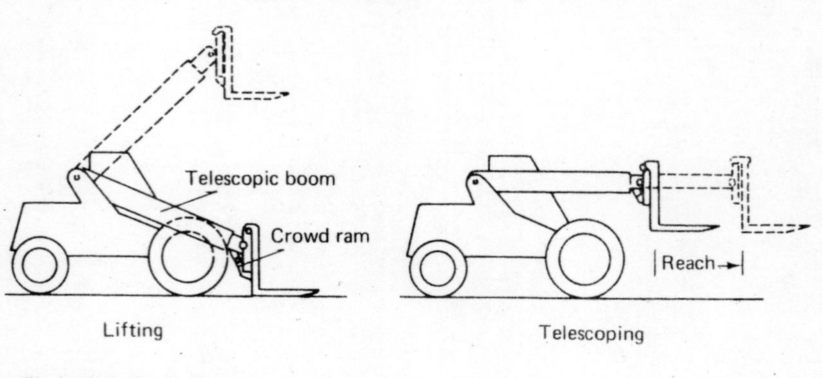

Figure 8.4 Boom forklift.

steering axle, which also takes weight off the rear wheels, so reducing traction.

The purpose-built forktruck is made in 'industrial' or 'rough terrain' versions. The industrial machine is designed for use on concrete or other hard, even surfaces, and is compact and highly manoeuvrable. The rough terrain truck is built of tractor chassis components, often incorporating four wheel drive, to enable it to work in field conditions. It can also be used in place of the industrial truck, but is larger and less manoeuvrable. In both types that use two wheel drive, the driving wheels are at the fork mast end, and the steering wheels at the rear.

The driving position faces the forks, and the controls are arranged for easy operation of the mast functions. The load counterweighting is incorporated in the rear end; in electric forklifts the battery pack is also used as the counterweight. The counterweight biases much of the fork-truck weight on to the rear steering axle when it is unladen. On soft ground this can cause traction problems as the undriven wheels sink, and there is insufficient weight on the driven ones to effect grip.

(h) Forklifts can be powered by either an internal combustion engine or an electric motor. The engine is normally fuelled by either diesel or liquefied propane gas (LPG), and power for the electric motor is supplied by storage batteries. The electric truck is restricted in its operations by the amount of power that can be stored in the batteries which are normally sized to give sufficient running time to cover a full 8 hour shift; they are then recharged from the mains through a transformer/rectifier over several hours. The battery cannot be charged rapidly, as this causes premature deterioration; it will still gradually deteriorate as a result of normal charging and discharging, but its life should be in excess of 5 years. An electric truck is cheaper to maintain and run than an engine driven one, and does not pollute the atmosphere of the building in which it works.

The engine-driven truck produces unpleasant and noxious fumes when working for long periods in enclosed buildings. Some are fitted with 'exhaust gas purifiers' to remove some of the lethal discharges like carbon monoxide. This does not, however, compensate for the oxygen consumed in an enclosed space. Most purifiers are only successful when kept up to operating temperature by prolonged engine use.

(i) The truck needs to reverse direction (shuttle) quickly, and preferably should have an infinite speed range up to its maximum. This precludes normal clutch and gearbox transmissions from engine-driven trucks; instead, oil-based systems are fitted, being either hydrokinetic (torque converter) or hydrostatic. The electric truck uses conventional d.c. motor control systems to effect speed variation and travel direction. On most modern trucks, the travel speed and direction are controlled from a single foot pedal, leaving the driver's hands free to operate the mast controls.

8.2.2.2 Forklift attachments

A wide variety of handling items other than forks can be fitted to the backplate, the two most common being as follows.

(a) Box tipper. Two types are used, forward tip where the forks hinge forward, or side tip where the backplate is in the form of a vertical turntable. Both systems use devices to hold the box on to the forks during tipping. The forward tip uses a top clamp arm, and the rotator uses a side arm to prevent the box from slipping sideways in conjunction with bottom deck boards to hold the base on to the forks. As the hinge point of the forward tipper is partway along the forks, boxes with a bottom deck cannot be tipped forwards.
(b) Buckets. Many types are made to handle soil, manure, sand or bulk crops. The bucket-mounting frame incorporates rams to tip it forward.

The hydraulic power to operated attachments may be taken from the device normally used for sideshift, or a further 'auxiliary services' valve spool can be added to the control bank.

8.2.2.3 Fork truck handling capacities

All trucks have a nominal load rating, for example, '1500 kg' or $2\frac{1}{2}$ tonne'. These capacities are conditional on how far along the forks this load is applied, this distance being termed the 'load centre' (figure 8.3). The normal load centre is either 500 mm or 600 mm; longer ones derate the truck (table 8.2).

Table 8.2

Capacity of a nominal 2270 kg forklift at various load centres

Load centre (mm)	*Capacity (kg)*
500	2270
600	2160
900	1730
1200	1500

The nominal capacity of a forklift will also be derated by the addition of other handling fittings. The weight of the fitting must be deducted from the nominal load capacity, to arrive at the useful load capacity. Forward-acting tippers seriously extend the load centre, especially in situations where the load is held in the box until it is fully tipped. Lateral stability can be affected by moving the load sideways, owing to sideshift or side tipping.

Using a shorter load centre does not automatically uprate the truck however, as its capacity is also governed by the strength and vertical lifting capacity of the mast.

Forklifts based on industrial models bear a 'rating plate', which defines the effects of altering the standard load centre; some manufactured purely for horticulture do not carry such data – neither by rating plate nor in the operator's handbook.

8.2.2.4 Dimensions

The important dimensional specifications are as follows.

(a) Closed mast height (figure 8.3).
(b) Width. Especially for a rough terrain unit, the overall width can be wider than the pallet, and thus determines the minimum width of aisle that the unit can be driven along.
(c) Length. Coupled with turning circle, this is sometimes termed the '90° stacking aisle width', the narrowest aisle that the truck can turn at right angles in to pick up a pallet from one side. For industrial trucks it is usually 3.6 m, but a rough terrain unit will require 5.4 m or more, especially if it is a four wheel drive vehicle in which the steering angle is limited.

8.2.2.5 Hand pallet truck (stillage truck)

These are used to move laden pallets within the packhouse or yard. The truck (figure 8.5) is raised on its wheels by either simple leverage from pushing the handle down, or hydraulic power generated by a pump fitted into the handle linkage.

The wheels, when retracted, protrude about 25 mm below the forks, so that the truck is low enough to enter the pallet base, but is still

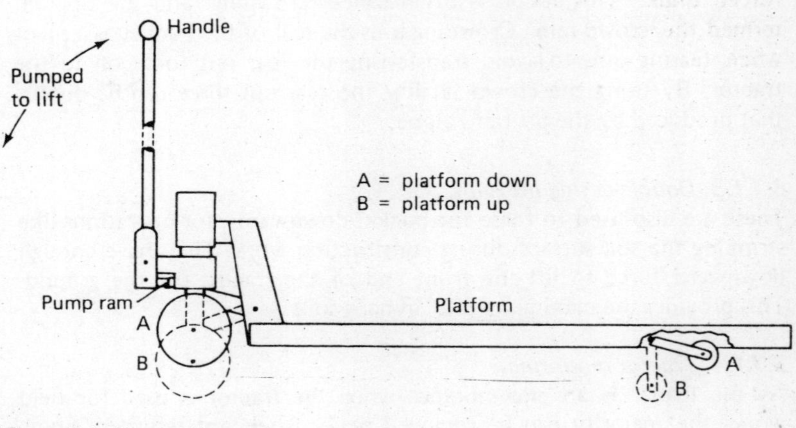

Figure 8.5 Hand pallet truck.

carried on them. The truck is best suited to pallets without a lower deck, as wheels can roll freely under. It will work on pallets with bottom deck boards if they are spaced to allow the fork rollers to pass through, or if the forks are long enough to protrude.

Electrically powered pallet trucks are used in large packhouses to eliminate the toil of hand propulsion. These can either resemble the hand type, but with electric drive, or have a small platform for the operator to ride on.

8.3 TRACTOR LOADERS

These are versatile handling attachments which can be fitted to standard tractors and a wide range of small horticultural and garden tractors.

8.3.1 Loader configurations

8.3.1.1 Basic loader

This uses hydraulic power only to lift the arms, which allows it to be powered directly from the tractor's trailer-tipping port without further valving. The bucket is held in its horizontal position by a latch, which is released to allow it to tip for emptying. Some buckets are weighted on the rear side of the pivot, to 'roll back' to horizontal when the load is discharged; others require to be reset.

8.3.1.2 Hydraulic bucket operation

The bucket is held by a ram which is extended for tipping and retracted for return; a special double-acting spool valve powers the ram in both directions. The tip rams can be built with sufficient strength for levering the bucket backwards to tear out hard material once the front edge is forced under. This action is often called 'crowding' and the tip ram termed the 'crowd ram'. Crowding uses the rear of the bucket as a pivot when tearing out, to avoid transferring the tear out force on to the tractor. By using the crowd facility the tear out force can be double that produced by the lift rams alone.

8.3.1.3 Double-acting lift rams

These are also used to force the bucket downwards, for operations like stripping the soil surface during construction work. They have enough downward force to lift the front end of the tractor off the ground. This provides the maximum 'bite' in hard soil.

8.3.1.4 Tractor mountings

As the loader is an encumbrance when the tractor is used for field work, the majority can be removed easily when not required. Some loaders are attached to the tractor on a 'quick fit' subframe, while others permit only the arms and bucket to be removed easily.

The extra weight put on to a tractor front axle by a loader often necessitates the fitting of power steering, if this is not already built into the tractor. The front tyres should also be uprated to carry the extra weight, and provide flotation on soft ground.

The loader transfers weight away from the tractor's rear driving axle which, together with the extra weight on the steering wheels, reduces traction ability. This can be compensated by adding rear weight, often in the form of a large concrete block on the three point linkage.

8.3.2 Loader specifications

When choosing a loader it is important to determine its weight capacity, lift height and forward reach.

(a) Lift weight capacity is dependent on both loader design and the tractor to which it is fitted. Like the fork truck, this capacity changes with the distance of the load ahead of the front axle. The capacity is quoted either at the bucket attachment pins or a small distance ahead of them.
(b) The lift height determines the height of container that can be filled. As the front edge of the bucket drops for tipping, the maximum clearance under a tipped bucket is the most useful measure of lift height.
(c) Forward reach is the horizontal distance between the tractor front and the bucket when fully raised. It is important when heaping loose material, like small coal or root vegetables, which roll back towards the tractor's front wheels from the heap top.

Lift and lowering times are also important when using a loader intensively, to avoid having to stop during each cycle for the bucket to ascend or descend. They are determined by both the flow capacity of the oil system, and the tractor's oil supply.

8.4 CONTINUOUS CONVEYING SYSTEMS

Material can be moved in a small but continuous stream by one of three methods, belt conveyor, auger or pneumatic.

8.4.1 Belt conveyors and elevators

The term 'conveyor' is normally reserved for belts that transport materials horizontally or on a slight incline, and 'elevator' for those that act vertically. The basis of both is a continuous band of either rubberised fabric belt or link chain.

8.4.1.1 Belting

Most 'rubber' conveyor belts are made of alternate layers of strong fabric and rubber. The fabric layers provide tensile strength and proof against stretching when the belt is under tension. The 'rubber' might be natural, but is more commonly synthetic or plastic which has been specially formulated for the belt's intended duty. The topmost ply is always rubber, and can possess such characteristics as abrasion resistance for handling coal and sand, or imperviousness and ease of cleaning for food processing. The surface texture can also be varied, and includes heavy patterning to prevent objects from slipping. In many simple conveyors the belt slides directly on wood or metal support sheeting. This requires the lower face to be wear resistant and have low frictional resistance.

The numbers and thicknesses of the fabric and rubber plies vary with specification, many light conveyors having only one ply of each. Thick, multi-ply belts are stiff and there will be limitations to the minimum diameter of pulley on which they can run. If these are ignored, the upper plies will crack prematurely and the lower ones might delaminate.

Elevator belting often requires a series of bars (flights) across its upper face, to prevent material rolling backwards. These might be wood or metal bars, bolted on to a flat belt, but are more commonly integral mouldings of the top ply rubber. Moulded flights can be thin strips up to 75 mm tall, set across the belt at intervals, or a series of low, closely spaced V-shaped bars (chevrons). Tall, transverse flights have three disadvantages.

(a) The flight tip generates sufficient speed when passing round a roller to throw material forwards and damage susceptible produce.
(b) Produce cannot be transferred on to the belt closer than the flight height, and produce landing on a thin flight edge is likely to be damaged.
(c) It is difficult to fit rollers against the flighted face to support the return strand of the belt. This is sometimes accomplished by leaving 25–50 mm of each edge of the belt unflighted, or gaps in the flights provide space for a thin wheel to run.

The chevron belt overcomes these difficulties, but is more expensive, heavier and needs larger-diameter end rollers. These latter factors tend to preclude its use on field machines.

8.4.1.2 Chains

These include roller and hook link drive chains with carrying slats mounted on them, or complete conveyors made of interlinked wire or plastic segments. Chain and slat systems can have either an entire surface of bars for carrying, or spaced transverse bars to push material along a flat sheet. Slat-carrying chains can support loads better than can rubber belt conveyors, and are used in, for example, moving floor bulk hoppers. The chains also provide stronger drive than belt and roller friction.

Interlinked conveyors are often used for draining washed materials or grading out fine soil and under-sized produced. Materials can be chosen for specific uses; for example, the food-processing industry uses conveyors formed of nylon segments, held together with stainless steel pins.

8.4.1.3 Conveyor drives

The only means of transmitting power to a belt is the frictional force between the belt under-surface and a metal roller. This force is governed by the coefficients of friction of the belt and roller surfaces, and the belt's pressure on the roller. Roller pressure is provided both by initial tension of the belt, and the tension generated by load resistance when the belt is running.

Often sufficient initial tension is provided by the weight of the return strand hanging down, either in one loop or between widely spaced support rollers. The drive roller should be at the discharge end of a conveyor or elevator to make full use of belt pressure generated by these means. If the roller is at the other end, the slack is transferred to between the roller and loading point, effectively reducing roller pressure as load increases.

Belting normally runs at speeds of 0.5–2 m/s so that a 200 mm drive roller will be rotating at 45–200 rpm. This requires heavy reduction from standard motor or engine speeds, which can be done with pulleys of large-diameter difference, or mechanical gearing. One electric drive incorporates a special motor and gearbox within the drive roller, so that its spindle is stationary, and its casing is the roller surface that rotates.

Belting has to be aligned with the conveyor frame, otherwise it runs towards one side and will either wear that edge or run off completely. Alignment adjustment is possible by mounting one of the end rollers in sliding bearing housings, so that the roller can be angled away from the run out. One common 'automatic' method uses a 'crowned' drum, which is of larger diameter in the centre than at each edge. The belt will always try to wander towards the larger diameter part of the roller, but when there is an equal amount either side, the side forces are balanced and the belt is held centrally.

8.4.1.4 Conveyor construction

Conveyors are often made to suit specific purposes, and there are several aspects to consider in their basic design. All are made of a deck unit which supports the belt or chain, a drive roller and an idler roller at the opposite end. The deck might be flat sheet on which the belt runs directly, or it might form a shallow trough to help contain the conveyed material. Running the belt on small idler rollers reduces frictional resistance. These rollers might be in three parts, the outer two angled upwards for troughing. Linked chain conveyors are normally flat as the transverse pins cannot flex. The deck unit can be mounted on legs which are adjustable for height, and castored to form a multi-use conveyor. Where it is necessary to alter the length of a run of conveyors, it is possible to mount the end of one above the next, 'piggyback' style, so that they can telescope.

The conveyor can have three basic purposes, transport, processing or inspection. Transport conveyors can be operated at high belt speed, 1.5–2 m/s, carrying a narrow band of material. Processing operations, such as draining and grading, require wide belt faces to spread material thinly. There must also be a facility to remove material that falls through the meshes, either by interposing a transverse catching conveyor, or allowing it to drop through the return strand as well, with a plough or loose link to eject material that holds back. Inspection conveyors run at 0.5–0.75 m/s, to enable visual analysis and picking off to take place. Width is normally limited to 600 mm for one-sided staffing, as this is the limit of comfortable reach. The height must suit the operator. Longitudinal guide boards are often fitted to form lanes for carrying each selection separately.

The elevator is normally an inclined belt conveyor with its lower end formed into an intake hopper. Most modern elevators use hydraulic rams for raising the upper end, and the ram of crop store loading machines can be controlled automatically, to keep the upper end close to the heap as it forms. Store loading elevators also distribute the crop laterally either by having a small horizontally swinging extension 'luffing boom' at the top end of the main conveyor, or by swinging the whole elevating conveyor. This latter has a specially wheeled chassis, whereby the outer wheels can turn sideways and allow the whole elevator to move in an arc across the floor. Small reversible motors drive these wheels to cause the elevator to traverse back and forth automatically.

8.4.2 Auger

This is based on an Archimedean screw (flight), rotating inside a tubular case, and will convey only free-flowing granular materials like dry peat, small coal or soil. In simple augers the flight runs freely in the tube without being supported by bearings. Often the flight edge uses the conveyor material as a lubricant against the tube wall, which can cause rapid wear when abrasive materials are being conveyed. Better augers have a flight supported by bearings to reduce the wear between flight and case. Problems can be encountered when conveying material with particles of the same size as the gap between the flight and case, which can lodge and stall the auger.

The auger can be used to convey or elevate, even running vertically, but the throughput declines as the angle increases (table 8.3).

Table 8.3

Throughput of a 150 mm auger on cereal grain

Angle (degrees)	0	15	30	45	60	70	90
Throughput (tonnes/h)	40	36	32	27	23	18	12

8.4.3 Pneumatic conveying

This uses a flow of moving air to blow material along tubes. The material must be powder or granular, and free flowing. The main advantage over the other methods is that it can convey material along tortuous paths, instead of only in straight lines. It does, however, absorb much higher power per unit of material shifted. For example, an auger lifting 5 tonnes of seed per hour, over 10 m, will use 1.0 kW; a blower 4.5 kW. The pneumatic conveyor also requires a more complex system for getting material into and out of the airstream.

Material is loaded into the airstream by either a venturi injector or a rotary valve. In the injector the airstream passes through a tapering orifice, and then is allowed to re-expand to the original pipe size. A partial vacuum is created in the zone of the minimum pipe diameter, which is used to draw in the material. The rotary valve is a compartmented wheel, mounted in a closely fitting housing. The base of the housing is connected to the conveyor pipe, so that material drops from each compartment into the airstream as the wheel rotates. The blades forming the boundary between each compartment prevent air from escaping.

Material can be removed from the airstream by allowing the flow to expand until its velocity has fallen below conveying velocity; this however requires a lot of space and can create dust. The most common method of material removal involves the use of cyclones, as described in section 7.6.3.

8.4.4 Roller conveyors

As this name implies, the conveyor bed is formed from a series of rollers. This type of conveyor can transport only rigid items, such as pallets, bins or crates; it is not suitable for loose material or soft packages, such as loosely filled sacks.

Free-running roller conveyors (often called 'gravity rollers') are the most common type found in horticulture, although powered rollers are used in certain applications. The gravity roller is made of a length of plastic or metal tube mounted on to a shaft by bearing races; by this means the shaft does not rotate and can be fixed to the supporting framework. Powered roller conveyors can use either rollers with solid shafts, supported in bearings on the frame and driven with chains and sprockets, or free-running rollers driven by a belt pressing against their undersides.

The gravity roller conveyor is used either horizontally, with the packages being pushed along, or on a slight slope, so that they move by gravity. The latter type is often used to accumulate packages from the end of a belt or chain conveyor. Many gravity roller units are simply constructed from two side members, with the rollers mounted between; this can then be placed on legs for permanent installation or propped on crates for use in temporary locations.

Bins or pallets for use on roller conveyors must have a bottom deck; if this is made of boards, they should be placed to run at right angles to the rollers, so that they span several rollers. If pallets are used with the bottom boards parallel to the rollers, the boards will be twisted as they roll across, and will eventually split or loosen.

Flat-bottomed packages, such as cardboard boxes, can be handled on wheeled roller conveyors; these have a number of wheels on each shaft, instead of the full roller.

8.5 MONO-RAILS

These are used for transport within buildings or other permanent structures. They fall into two basic types.

8.5.1 High-level mono-rails

8.5.1.1 Non-powered

This type is commonly seen in glasshouses and consists of a trolley which is hung from the high-level heating pipes. It is used to transport items of produce along the bays to one end or to a cross access road. If the pipe is reasonably level, the trolley will run to its destination after a single hard push.

The trolley hangs on the pipe by a simple wheeled cradle, which can be hooked or unhooked easily (figure 8.6), but this does require the pipe to be mounted on the type of supports shown in figure 8.6. The load capacity will depend on the pipe strength and its supports, but most can take at least 10 kg.

This system requires the loading to be synchronised with the unloading because the empty trolley has to be returned along the same pipe run.

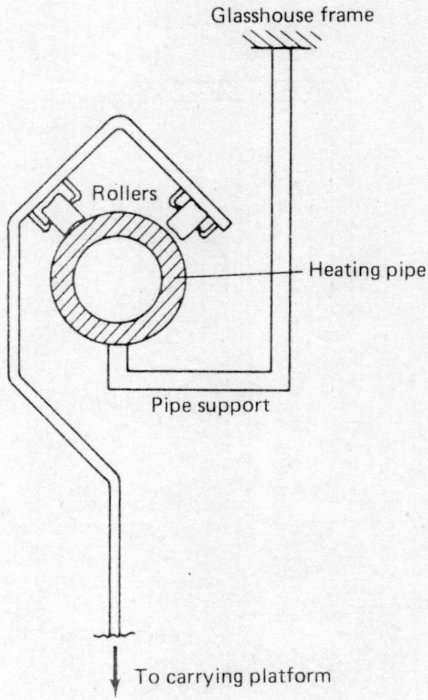

Figure 8.6 Non-powered mono-rail.

8.5.1.2 Powered

This system is based on a continuous loop of track which carries a number of carriages linked into an endless chain.

The track is normally made of a specially shaped rolled metal section (figure 8.7(a)), although light-weight systems have been based on pipe (figure 8.7(b)). The driving chain runs beneath the track; it has to be capable of flexing both vertically and horizontally (bi-planar); guide sprockets are placed at each change of direction. A power unit is sited at a convenient place alon the chain run; if the system is arranged so that only one leg is used for the loaded trollies, it is best to site the power unit immediately after the unloading point.

The track has to form a continuous loop, although this can be in a fairly complex shape in order to give access to all the handling points. Systems with long 'return' legs can prove relatively expensive, because the same track will be used for this portion as for the loaded sections.

This system allows asynchronised working, because any carriers not filled or emptied can be left to go round again. The load capacity can range from a few kg to several tonnes per carrier.

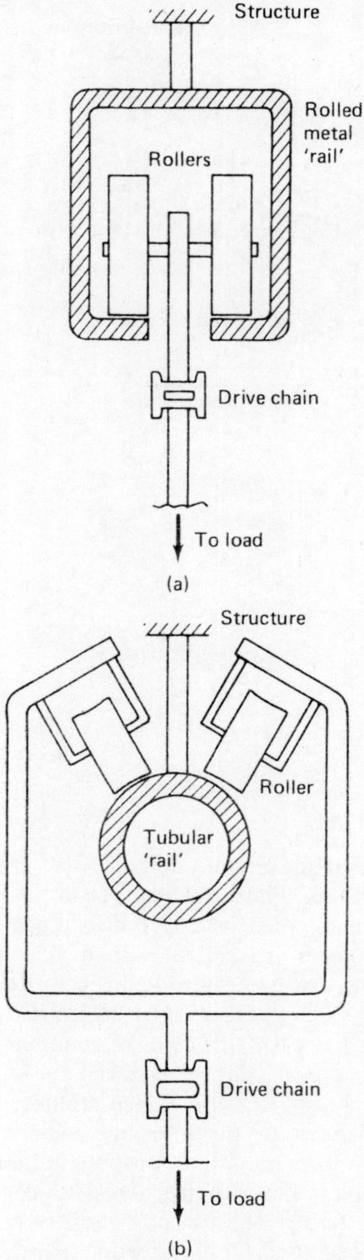

Figure 8.7 Types of powered overhead mono-rail: (a) rolled section, (b) pipe track.

8.5.2 Ground-level mono-rails

These operate on a track formed from special lengths of girder, which are assembled to run along the ground like a single rail line. The track girder section is normally 3–4 m long, and 250–300 mm high, with a broad top on which the train's support wheels run. Opposing pairs of horizontal wheels run either side of the lower part of the girder to hold the train upright.

The girder sections have adjustable stand feet, to allow the track to be laid over uneven ground. The joints between each section accept angular misalignment, to allow the track to be curved.

The load is carried on small trollies, propelled by a motor trolley; the train is normally controlled by a person walking alongside.

This type of mono-rail is suited to transporting heavy material between different sites on the nursery; for example, loaded bulb-forcing trays from the standing ground to the glasshouse. The track is portable and easily assembled, so that its route can be changed to follow the progress of the work. The system is normally laid open-ended, that is, the empty train reverses to the filling point.

APPENDIX A: METRIC CONVERSION FACTORS

Length	1 mm	=	0.0394 inches
		=	0.1 centimetres
	1 m	=	3.28 feet
		=	1.09 yards
	1 km	=	1093.6 yards
		=	0.62 miles
Area	1 m^2	=	10.76 square feet
		=	1.196 square yards
	1 ha	=	10 000.0 m^2
		=	2.47 acres
		=	11 956 square yards
Volume	1 litre (l)	=	1.76 Imperial pints
		=	2.11 U.S. pints (1 U.S. liquid pint = 0.859 U.S. dry pints)
		=	0.22 Imperial gallons
		=	0.26 U.S. gallons
	1 ml	=	1.0 cubic centimetres
		=	0.0352 fluid ounces
	1 m^3	=	35.31 cubic feet
		=	1.31 cubic yards
Mass	1 g	=	0.035 ounces
	1 kg	=	2.204 lb
		=	35.26 ounces
	1 tonne (t)	=	2204.6 lb
		=	0.984 Imperial (long) tons
		=	1.102 U.S. (short) tons (1 short ton = 2000 lb)
		=	19.6 Imperial hundredweights
		=	22.05 U.S. cwt

Force	1 Newton (N)	=	0.101 kilogrammes force
		=	0.225 lb force
	1 kN	=	0.10 tons force
Pressure	1 N/m^2 (Pa)	=	0.10 mm water gauge
	1 kPa	=	4.02 inches water gauge
		=	0.144 lbf/square inch (psi)
		=	0.01 bar
		=	0.0102 kgf/cm^2
		=	0.102 m water gauge
	1 MPa	=	144 lbf/square inch
Power	1 kW	=	1.34 horsepower
Energy	1 kiloJoule (kJ)	=	0.948 British thermal units (Btu)
		=	0.238 kilocalories
	1 MJ	=	0.278 kWh
Volume flow	1 litre/s	=	13.2 Imperial gallons/min (gpm)
		=	791.9 Imperial gallons/h
		=	15.8 U.S. gallons/min
		=	949.9 U.S. gallons/h
	1 m^3/s	=	35.31 ft^3/s (cusec)
		=	2118.6 ft^3/min (cfm)
Proportions	1 g/ha	=	0.014 oz/acre
	1 kg/m^2	=	1.85 lb/yd^2
		=	0.205 lb/ft^2
	1 kg/ha	=	0.0079 Imperial cwt/acre
		=	0.892 lb/acre
		=	0.0029 oz/yd^2
	1 ml/m^2	=	0.029 fl oz/yd^2
		=	7.02 Imperial pints/acre
	1 litre/ha	=	0.089 Imperial gallons/acre
		=	0.107 U.S. gallons/acre
	1 ml/ha	=	0.0143 fl oz/acre
	1 ml/litre	=	0.16 fl oz/gal
	1 g/litre	=	0.159 oz/Imperial gallon
		=	0.130 oz/U.S. gallon
	1 t/ha	=	7.96 Imperial cwt/acre
		=	8.92 U.S. cwt/acre
		=	0.398 Imperial (long) tons/acre
		=	0.446 U.S. (short) tons/acre

100 seeds per 10 g	=	3.5 oz per 1000 seeds
1 Pa/m head loss	=	0.0044 psi/100 ft
	=	0.010 ft water gauge/100 ft

APPENDIX B: FURTHER READING

Horticultural Skills Training Guides, Agricultural Training Board, Bourne House, Beckenham, Kent, BR3 4PB.

Health and Safety in Agriculture (guides and regulations for U.K. horticulture), Health and Safety Executive, Baynards House, 1 Chepstow Place, London W2Z, 4TF.

C. Culpin, *Farm Machinery*, Crosby-Lockwood, St Albans, 10th revised edn, 1981.

C. Culpin, *Profitable Farm Mechanisation*, Crosby-Lockwood, St Albans, 3rd revised edn, 1975.

Pumping and Irrigation, The Electricity Council, Farm Electric Centre, Stoneleigh, Warwickshire, CV8 2LS, revised edn, 1983.

B. Davis, D. Eagle and B. Finney, *Soil Management*, Farming Press, Ipswich, 4th revised edn, 1982.

C. Bishop and W. Maunder, *Potato Mechanisation and Storage*, Farming Press, Ipswich, 1980.

J. Robertson, *Mechanising Vegetable Production*, Farming Press, Ipswich, 2nd revised edn, 1978.

B. Bell, *Farm Workshop*, Farming Press, Ipswich, 1981.

M. Neidle, *Electrical Installation Technology*, Newnes-Butterworths, London, 3rd revised edn, 1982.

INDEX